RECENT PROGRESS IN MICROCALORIMETRY

Recent Progress in MICROCALORIMETRY

by

E. CALVET

**Professor at the Faculty of Sciences, Marseilles
and Director of the Institute of Microcalorimetry
and Thermogenesis**

and

H. PRAT

**Professor at the Faculty of Sciences, Marseilles
and the University of Montreal**

Edited and translated from the French by

H. A. Skinner

**Senior Lecturer in Physical Chemistry
University of Manchester**

PERGAMON PRESS

OXFORD · LONDON · NEW YORK · PARIS

1963

PERGAMON PRESS LTD.
Headington Hill Hall, Oxford
4 & 5 Fitzroy Square, London W. 1.

PERGAMON PRESS INC.
122 East 55th Street, New York 22, N.Y.

GAUTHIER VILLARS ED.
55 Quai des Grands-Augustins, Paris 6

PERGAMON PRESS G.m.b.H.
Kaiserstrasse 75, Frankfurt am Main

Distributed in the Western Hemisphere by
THE MACMILLAN COMPANY · NEW YORK
pursuant to a special arrangement with
PERGAMON PRESS LIMITED

Translated in part from the French edition
"*Récents Progrès en Microcalorimétrie*"
Dunod, Paris, 1958

Library of Congress Card No. 62-9175

PRINTED IN HUNGARY

CONTENTS

FOREWORD

AFTER the publication of our two books on the subject of Microcalorimetry*, several English-speaking colleagues suggested that it would be helpful to issue an English translation. As a great deal of new work has been completed in the five years since our first book appeared it seemed preferable to produce a completely new and revised book rather than a mere translation from the French editions. Furthermore, we decided it would be most practical to prepare this of a size intermediate between that of the treatise of 1956 and the monograph of 1958.

For this task we needed the assistance of someone able to re-mould our writings into good English and fully appreciative of the object and scope of our microcalorimetric studies. We were fortunate in gaining the help of our colleague and friend Dr. Skinner, of the University of Manchester, the President of the Commission of Experimental Thermochemistry of the International Union of Pure and Applied Chemistry. For his valuable help, we express our most grateful thanks.

E. CALVET and H. PRAT

Marseilles

* *Microcalorimétrie: applications physico-chimiques et biologiques.* Masson Ed., Paris, 1956.

Récents progrès en Microcalorimétrie, Dunod Ed., Paris, 1958.

INTRODUCTION

The interest of microcalorimetric studies

In general, all phenomena of a physical, chemical, or biological nature are accompanied by thermal effects, the precise magnitude of which may be of considerable practical and theoretical importance. Such measurements provide the fundamental data upon which calculations of heat exchange in industrial processes depend; in pure research, they may give valuable information concerning the stability and structure of molecules, in predicting the course of reactions, and the conditions of chemical equilibria, etc.

Thermodynamic studies of this type aim to measure the total quantity of heat, Q, produced by a given transformation, independent of the time factor.

The apparatus and microcalorimetric techniques which we shall describe diverge somewhat from classical procedures. They were begun by A. Tian in his studies during the period **1923–1933**. Since 1932, we have continued to develop Tian's calorimeter, and to extend its applications. At the present time, we are constructing very sensitive microcalorimeters capable of recording energy outputs of the order of a microwatt; they can be used to study slow processes of almost unlimited duration with remarkable fidelity, and furthermore they are easy to operate. The apparatus does not measure the quantity of heat produced directly, but records the calorific power*, W, produced

* The calorific power, W, is the *rate* of heat production $= \frac{dQ}{dt}$

at each instant, t, by the process under investigation. The experimental observations are recorded as a continuous curve, which, after minor adjustments, is a correct representation of W as a function of time: this curve of

$$W = f(t) \tag{1}$$

we refer to as the *thermokinetic* curve of the process being studied.

To obtain the quantity of heat, Q, produced between times t_1 and t_2, one simply measures the area under the curve (1) for the time interval $(t_1 - t_2)$; alternatively, the corresponding area under the directly recorded curve can be measured, subject to a small correction easy to estimate.

The thermokinetic curve (1) not only provides a measure of the heat evolution, Q, in a given time interval, but also gives an accurate representation of the *kinetics* of the process.

We shall describe several studies which have been made either in our laboratory at the Faculty of Science, Marseilles, or in the laboratory of Professor H. Prat at the University of Montreal, Canada.

It would appear to be generally useful as a preliminary study in a wide variety of research problems to obtain the thermogram of the phenomenon under investigation, in order to disclose the various factors influencing the phenomenon. One can foresee the microcalorimeter becoming a rather commonly used instrument in research laboratories within a period of 10–20 years from now.

PART 1

APPARATUS AND MICROCALORIMETRIC METHODS

by E. CALVET

CHAPTER I

DIFFERENT TYPES OF CALORIMETERS

1. The Special Features of our Microcalorimetric Apparatus

A calorimeter consists essentially of a container in which the thermal phenomena under investigation are carried out and studied. In general, this container is placed in a cavity, the walls of which are kept at constant temperature (the 'constant temperature environment' calorimeter), or, alternatively, at a temperature which can be varied at will (the 'adiabatic' calorimeter). We shall refer to the walls of the calorimeter container as the "*internal boundary*", and to the walls of the surrounding cavity as the "*external boundary*".

Heat liberated in the calorimeter container raises the temperature of the internal boundary, and results in an exchange of heat between the two boundaries. In the adiabatic type of calorimeter, this heat exchange is reduced to a minimum; by contrast, the heat exchange is fundamental to our microcalorimetric method.

The precise evaluation of the flow of heat, Φ, passing at each instant from the internal to the external boundary, is simple to make only when the temperatures θ_i and θ_e of the walls defining the internal and external boundaries are uniform. In this ideal case, if the temperature difference $\theta_i - \theta_e$ is sufficiently small, the relation

$$\Phi = p(\theta_i - \theta_e) \tag{2}$$

follows from Newton's law of cooling, where p is the thermal conductivity of the medium between the boundaries.

However, although it is not difficult to keep the external boundary walls at uniform temperature, the same does not apply to the internal boundary. The heat developed in the calorimeter container is not uniformly produced within the vessel, and even if the walls are of good-conducting metal, the wall temperature is unlikely to be uniform. It is normal practice to stir the contents of the calorimetric vessel in order to achieve temperature uniformity; but this remedy was not open to us, because the heat of mechanical stirring would often exceed the heat developed by the processes under investigation.

As we shall see later on, although the temperature difference between the two boundaries may be experimentally badly defined, this need not apply to the flow of heat, Φ, which can be measured with very high precision by means of a battery of thermocouples judiciously disposed between the two boundaries. Hence there is no need for concern about the lack of temperature uniformity, nor for stirring the calorimeter contents.

By integrating the graph of heat flow from the calorimeter, Φ, as a function of time, i.e.

$$\Phi = g(t) \tag{3}$$

one can measure the total quantity of heat liberated. We shall show that the thermokinetic curve, $W = f(t)$, may be deduced from (3). This is quite easily done for very slow thermal processes.

2. Classification of Calorimeters

Different types of calorimeter can be conveniently classified according to the magnitude of the thermal conductivity, p, in equation (2):

TYPE 1: *p = 0, or very small: adiabatic calorimeters.*

If $p = 0$, i. e., the two boundaries are separated by a perfect insulator, the calorimeter is said to be *adiabatic* (no heat exchange with environment).

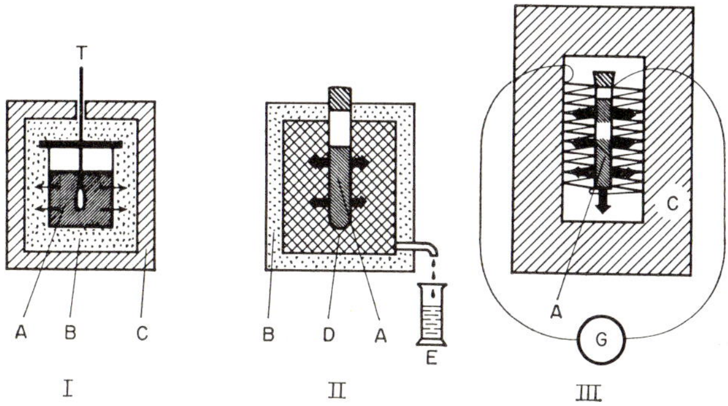

FIG. 1.1. Different types of calorimeters.

A (fine lines) = zone of heat production, inside the calorimeter container.

B (dots) = insulating gap.

C (broad lines) = external jacket, good conductor of heat.

D (crossed lies) = ice, or some other easily melted substance.

The arrows show the leakage of heat: the thickness of the arrows is representative of the magnitude of the leakage in each case.

I. *Adiabatic calorimeter*—Most of the heat produced is retained by the calorimeter container, causing its temperature to rise. The temperature change is measured by a suitable type of thermometer.

II. *Isothermal calorimeter*—All the heat produced in *A* is adsorbed in melting a certain weight of solid in *D*. The weight of liquid collected in the test tube *E* is proportional to the heat produced.

III. *Tian-Calvet calorimeter*—Most of the heat produced in *A* is *conducted* to the external jacket *C* by the bank of thermocouples surrounding the calorimeter container. The e.m.f. produced in the thermoelectric pile (i.e. the bank of thermocouples) is proportional to the calorific power transferred.

In these conditions, *all* the heat, Q, produced in an experiment is used in raising the temperature of the calorimeter. Thus, if the rise in temperature is θ,

$$Q = C\theta \tag{4}$$

where C is the heat capacity of the calorimeter.

This type of calorimeter responds to the *total* heat developed during an experiment, which includes the heat of reaction, of stirring, of evaporation, etc., and for this reason we shall refer to it as an "*integrating*" apparatus. A thermometer is required to record the temperature change, θ. The heat capacity, C, is measured by a calibration experiment in which a known quantity of heat, Q_0, is produced in the calorimeter, causing a temperature change, θ_0.

Obviously, by plotting Q against t, one could derive the calorific power, W, produced each instant, since

$$W = \frac{dQ}{dt} \tag{5}$$

Berthelot's calorimeter (shown schematically in Fig. 1.1) allows appreciable heat leakage from the calorimeter container, and is only approximately adiabatic. The leakages are difficult to evaluate correctly, detracting from the accuracy attainable with this type of calorimeter. Moreover, it is essential to stir the calorimeter contents, which limits its usefulness when required for the measurement of small heat outputs. The apparatus is not very sensitive, and can only be used in experiments of short duration.

The best types of adiabatic calorimeter are those in which the temperature of the external boundary is continuously adjusted to match that of the internal boundary, so that the net heat transfer between the two boundaries is zero. This result is achieved by automatically varying the temperature of the outer jacket to follow that of the inner cell, using a regulator which responds to any difference of temperature between the two boundaries. The idea was proposed by the French scientist, C. C. Person (1849), and first made use of practically by T.W. Richards (1905), since when it has been widely used.

Two disadvantages of the adiabatic type of calorimeter are that it has relatively large thermal inertia, and that the temperature of the calorimeter changes during the course of an experiment—for this latter reason it is not satisfactory for kinetic studies on chemical or biological transformations.

TYPE 2: *p very large: isothermal calorimeters.*

When the thermal conductivity p of the medium separating the two boundaries is very large, all the heat produced in the calorimeter container is rapidly transferred to the outer jacket. The temperature difference $(\theta_i - \theta_e)$ of equation (2) is thus kept very small, and it would be difficult to measure accurately; however, this is unimportant, since we dispense with the thermometer as recording instrument in this type of calorimeter and measure, not the temperature, but some quantity proportional to the heat flux Φ from the calorimeter.

The Bunsen ice-calorimeter provides one of the most perfect examples of a truly isothermal calorimeter*, and is shown diagrammatically in Fig. 1.1, II.

Another apparatus, sometimes used, is the semi-isothermal calorimeter of Junkers (1907). This depends on water-cooling to absorb the heat produced in the calorimeter container. The cooling water passes through a series of coaxial cylindrical passageways (in a labyrinth), and the temperature difference between the water on entering and leaving the labyrinth is measured.

For our purposes the main objection to the isothermal and the semi-isothermal types of calorimeter is that their thermal inertia is too large to allow them to be used as

* In the Bunsen ice-calorimeter, however, p is small.

oscillographs; the melting of ice in one case, and the rate of temperature equilibration by the cooling water in the other, occur too slowly.

TYPE 3: *Tian's conduction calorimeter, and Calvet's modification of it.*

Tian's microcalorimeter, with compensation by the Peltier and Joule effects, which we have gradually transformed into a differential microcalorimeter, also with compensation, does not belong strictly to either of the two above categories. We measure the heat flux from one boundary to the other, representing the greater part of the thermal power generated in the calorimeter container. The latter undergoes slight variations in temperature, which are, however, essential for the flow of heat to occur.

The main advantages of the conduction type of calorimeter are as follows:

1. The temperature changes of the calorimeter container are very small: for example, they may be less than 10^{-4} °C (corresponding to a deviation of the order 10–20 cm on the recorded curve). The calorimeter is therefore very well suited for the study of chemical or biological kinetics.
2. The calorimeter can be operated by a null-method, in which the heat liberated is compensated by use of the Peltier effect. It then behaves essentially as an isothermal calorimeter.

The early calorimeter of A. Tian resembled that shown schematically in Fig. 1.1, III (except for an additional thermoelectric pile for compensation by the Peltier effect), and possessed these advantages allied with high sensitivity. The fidelity of the apparatus was sufficient to allow study of processes of duration of several hours—even as long as one day.

Tian's early design of calorimeter has been tested over a period of years, and the experience gained thereby has enabled us to make substantial improvements to the original design, and operation, of the calorimeter. Chief among these are the following:

(a) *Correct measurement of the heat flow from the calorimeter, without the necessity for stirring the contents.*

The calorimeter container (cylindrical in shape, with diameter small relative to height) is surrounded by a sufficiently large number of *identical* thermocouples, *very regularly arranged.* If these conditions are met, and the thermoelectric pile is carefully made, the e.m.f. produced is then proportional to the heat flow from the internal boundary, irrespective of temperature variations over the boundary surface. Hence, we can dispense with the need for stirring the calorimeter contents—which is an immense gain when slow reactions of low heat output are being studied.

(b) *The use of twin calorimeters, connected in opposition. The joining together of microcalorimetric elements in series, opposition, etc.*

We have found that two identical Tian microcalorimetric cells, symmetrically placed in the same thermostat, with their thermoelectric piles connected in *opposition,* constitute a twin calorimetric system which is free from fluctuations in the zero-reading and unaffected by variations in the thermostat temperature.

Denoting the poles of the thermoelectric detector piles attaching to the external and internal boundary walls by E and I respectively, connexion in *opposition* joins I_1 and I_2, or alternatively E_1 and E_2, leaving the free poles to connect to the galvanometer.

An alternative arrangement of the cells, the addition, or *series* arrangement, is obtained by joining I_1 and E_2, or I_2

and E_1. We have found this arrangement useful for measuring heats of vaporization, although in practice, 4 microcalorimetric cells were used with two pairs of cells (series-connected) joined in opposition.

(c) *Compensation in the laboratory-element, coupled with the differential arrangement.*

In most differential calorimetric systems, the heat liberated in one of the twin elements is reproduced by electrical heating in the second, so that no e.m.f. is recorded by a thermoelectric pile connecting the two elements. The procedure is acceptable, in principle, if the two elements are identical: in practice, however, it is almost impossible to achieve exactly the same thermal leakage from the internal boundaries of the twin elements, even should they be heated by identical sources—and usually the sources of heat are different.

Accordingly we have adopted a different compensation procedure: in our differential calorimeter one of the twin elements acts as *reference-element,* and is maintained throughout at the temperature of the thermostat, whilst in the second or *laboratory-element,* the heat of the process under investigation is internally compensated by Peltier cooling (if exothermic), or by Joule heating (if endothermic).

(d) *Approximate compensation.*

For precise thermochemical measurements it is not necessary, nor is it desirable, to compensate *totally* the heat of the process in the laboratory-element. The complexity of the automatic devices needed to effect total compensation would, in any case, probably introduce additional sources of error.

We prefer to compensate from 90—95% of the total heat produced (the compensated heat can then be measured with high precision), and to measure the non-compensated heat from the curves recording the heat flux from the laboratory element. This latter can be evaluated within error limits of less than 1%, so that the overall accuracy in measuring the total heat is $> 99.95\%$.

For thermokinetic studies, it is not usually necessary to apply a compensation technique. The heat developed is directly recorded within error limits of 1%.

(e) *The transformation of lateral thermal perturbations into vertical perturbations, equally divided between the two microcalorimetric elements.*

This is discussed in detail later on p. 28

(f) *The suppression of parasitic e.m.f.'s.*

We have been unable to eliminate completely the parasitic e.m.f.'s which are caused by the non-homogeneity of the wires connecting the thermoelectric piles to the galvanometer and switches.

(g) *Operation of the microcalorimeter at high and at low temperatures.*

We have now constructed calorimeters suitable for studies at high temperatures (up to 1000 °C), and others suited to studies at low temperatures. These operate on essentially the same principles as those designed for normal temperature working, but naturally differ in the details of construction.

CHAPTER 2

THE MEASUREMENT OF THE HEAT FLOW BETWEEN THE CALORIMETER AND JACKET BOUNDARIES BY MEANS OF A THERMOELECTRIC PILE

1. Some Principles of Thermoelectricity

(a) Thermoelectric electromotive force (Seebeck effect)

Figure 2.1 is a schematic representation of a thermoelectric couple made up of the parts $A/B/A$, where A and B are two different conductors

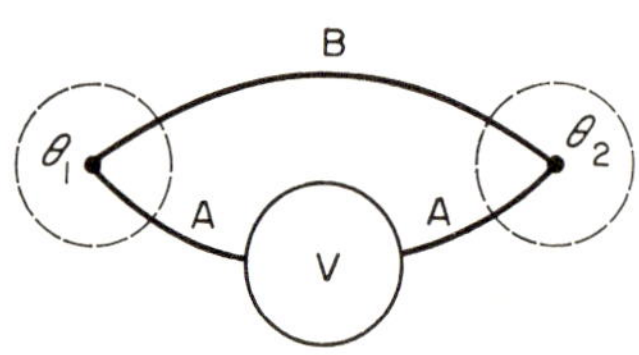

Fig. 2.1. Diagram of a thermoelectric couple.

If the two soldered junctions A/B and B/A are at different temperatures θ_1 and θ_2, a thermal e.m.f., E, results which is considered to be positive, if, on closed circuit, the current flows from B to A across the hot junction. The magnitude of E depends on the nature of the conductors A and B, as well as the temperatures θ_1 and θ_2.

Figure 2.2 represents a thermoelectric pile composed of n thermocouples joined in series with odd- and even-numbered junctions held at temperatures θ_1 and θ_2 respectively. The total e.m.f. of the pile is nE.

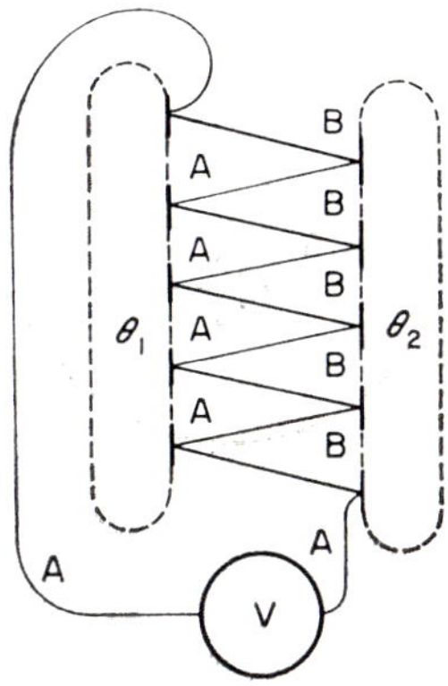

FIG. 2.2. Diagram of a thermoelectric pile

The *thermoelectric power* ε is, by definition

$$\varepsilon = \frac{dE}{d\theta} \tag{6}$$

As the temperature gradients $(\theta_i - \theta_e)$ occurring in our microcalorimeters are always small, the thermoelectric power ε is virtually constant over these gradients. Hence, the e.m.f. of a pile containing n couples can be expressed in the form:

$$E = n\varepsilon\,(\theta_i - \theta_e) \tag{7}$$

(b) The Peltier effect

Let us replace the voltmeter in Fig. 2.1 by a battery producing a current i in the circuit, and suppose that before passing the current the whole system was at a uniform temperature θ (Fig. 2.3).

During the passage of the current one of the junctions becomes warm whilst the other becomes cold. On changing the direction of the current, the hot junction cools and the cold junction heats up: this phenomenon is known as the *Peltier effect.*

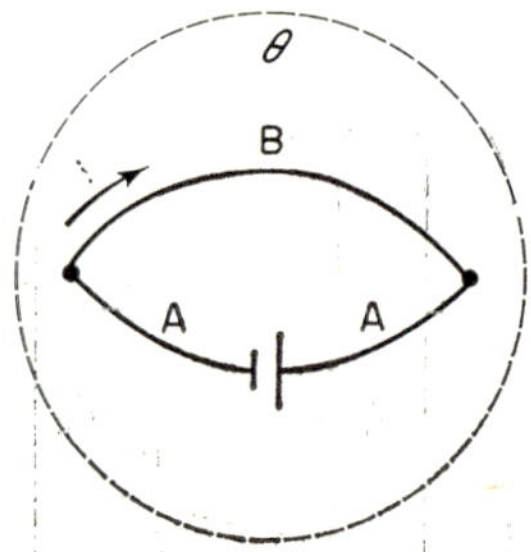

FIG. 2.3. The Peltier Effect. On passing a current through the circuit, one of the metal junctions A/B cools, whilst the other warms up, both being originally at the same temperature, θ.

The calorific power produced at each junction is proportional to the current, i. Thus,

$$W = Pi \tag{8}$$

where P is the Peltier coefficient for the couple A/B at temperature θ.

In practice, the Peltier effect, localized at the junctions, is superimposed on the *Joule effect* produced throughout the circuit in consequence of the passage of the current, i. The Joule effect, of the form Ri^2, is always positive, whilst the sign of the Peltier effect depends upon the direction of i. The heat produced by the Joule effect is transmitted in part to the junctions, so that they heat up, as if they had an effective resistance ϱ. The heat flow at the junctions is therefore given by

$$W = \pm Pi + \varrho i^2 \tag{9}$$

The value of ϱ is obtained by finding the current, I, which annuls the heat flow at the cold junction, when

$$0 = -PI + \varrho I^2, \text{ so that } I = \frac{P}{\varrho} \tag{10}$$

Substitution of this value of ϱ into equation (9) gives for the Peltier cooling effect

$$W = Pi\left(1 - \frac{i}{I}\right) \tag{11}$$

We should point out that the value of ϱ is very small in our most recent microcalorimeters.

(c) The choice of materials for making the thermocouples

Not only the thermoelectric power, but also the thermal and electric conductivity has to be borne in mind when considering the suitability of different types of thermocouple for microcalorimetric purposes. This question will be examined in Chapter 6, in which the sensitivity of a microcalorimeter is discussed; meanwhile, it is sufficient to report that after using iron-constantan couples for a long time, we are now making use of chromel-constantan couples, which are more efficient.

2. The Use of Thermoelectric Piles as Thermometers

Although a thermoelectric pile, in conjunction with a galvanometer, is an extremely sensitive device for detecting small temperature differences (of the order of one-millionth of a degree), we have not chosen to use them as thermometers for measuring the temperature difference between the two boundaries of a calorimeter, for the reasons below:

(a) The temperature of the internal boundary is not well-defined

The external boundary is at the temperature of the thermostat in which it is situated, and may be stable to *ca.* 1/1000th of a degree, which is the limit, in practice, to which a thermostat can be kept constant. On the other hand, the internal boundary is not at a uniform temperature, since the contents of the calorimeter container are in a state of "thermal disequilibrium". The thermometer sensitivity to 10^{-6} degree is therefore quite meaningless in this situation.

(b) The temperature of the thermoelectric junctions is not the same as that of the boundary wall to which they are attached

The thermoelectric junctions surrounding the metallic calorimeter container must necessarily be insulated from it, so there is a temperature gradient across the thickness of the insulating material. Furthermore, the conducting wires of the thermocouples carry the heat from the hot to the cold junctions, so that the temperature gradient across the junctions is always less than that between the internal and external boundaries.

Accordingly, we do not attempt to use the thermoelectric piles as thermometers, but only for recording the *heat flux*.

3. The Use of Thermoelectric Piles for Measuring Constant Flows of Heat

We shall review rather quickly some of the "classical" methods of measuring heat flux, paying particular attention to the defects of these procedures: this will lead naturally to the new ideas behind the construction of our apparatus.

(a) Single calorimetric systems

Consider a single calorimeter container, having one or more thermoelectric junctions attached to the external boundary wall, the others being kept at constant temperature, T_0. The thermoelectric pile is connected to a galvanometer.

When a source of constant heat power, W, is placed in the calorimeter container, the temperature of the container rises, and reaches a steady value, T, when the thermal leakage balances the heat input, W. The galvanometer deflection in the thermoelectric pile circuit indicates a current flow of intensity, i.

The magnitude of W may be determined by a substitution method with electrical energy. The original heat source

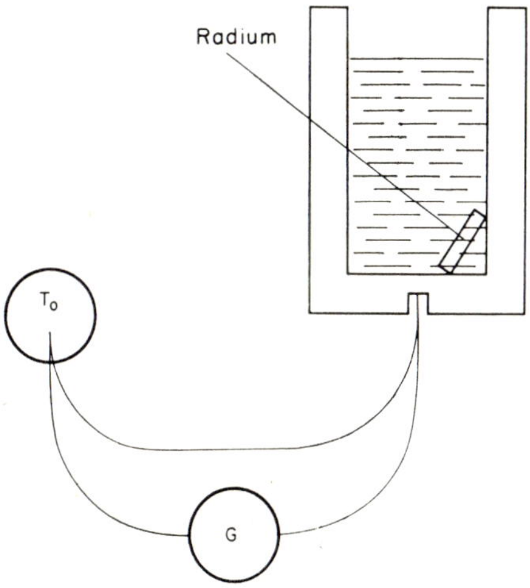

FIG. 2.4. A "classical" method of measuring the flow of heat from a calorimeter. The method is defective.

is replaced by an electrical resistance carrying a constant current and producing, by the Joule effect, a known heat power, W'. In the event that the galvanometer deflection indicates the same current intensity, i, as previously, it is assumed that $W = W'$.

This method presumes that *different* sources of equal heat power when put in the calorimeter will lead to the *same* temperature distribution. However, experiments have shown that the temperature distribution inside, and on the walls of the calorimeter container, depends upon both the position and the nature of the heat source.

(b) Differential or twin calorimetric systems

Several investigators have tried to improve the preceding method by carrying out the heating experiment and the electrical calibration experiment simultaneously. For this, two identical calorimeters are used, connected by thermocouples attaching to corresponding points on the

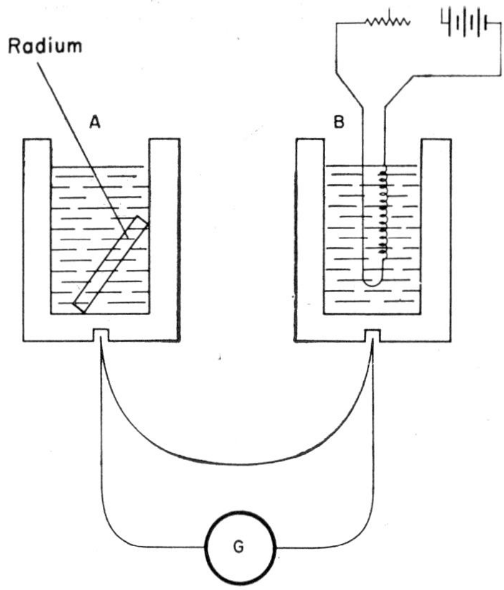

FIG. 2.5. The so-called "differential" arrangement. The heat flux from calorimeter A is compared with that from calorimeter B. This method is also not entirely sound.

two calorimeter containers (Fig. 2.5). Heat is developed in both calorimeters simultaneously in such a manner that there is no deflection of the galvanometer. The heat power, W, developed in one of the calorimeters is then identified with the known heat power, W', produced electrically in the other calorimeter.

This procedure is not strictly a *compensation method*, since both calorimeters warm up during the experiment. It is correct to compare the heat flows from the twin calorimeters only in so far as they are truly identical, and in our view it would be almost impossible to achieve identity to a degree comparable with the precision we have sought in our calorimeters. Furthermore, the same objections to those already mentioned under (a) in respect of the single calorimetric system apply to the twin calorimetric system.

4. The Correct Measurement of Heat Flow from a Calorimeter Container

To eliminate the disadvantages of the above procedures, two possible solutions present themselves:

1. The calorimeter container is embedded in a thick block of good-conducting metal, to achieve temperature uniformity. This method has been successfully used to measure the heat evolution by a radioactive substance, which was placed in the centre of lead sphere, in which the total radiation emitted was absorbed. The response of a thermocouple in contact with metal wall was found to be independent of the position of contact on the wall. This type of calorimeter, however, has a large thermal inertia and would not be suitable for use as an oscillograph.

2. The solution we have adopted is to completely surround the wall of the calorimeter container (i.e. the internal boundary) with *identical thermoelectric junctions, each separated from one another by equal intervals. The couples must be affixed normal to the calorimeter wall, and there must be a sufficient number of them* so that each covers only a minute portion of the total boundary surface. Assuming these conditions are met, each individual junction may be considered to be at a uniform temperature, θ_i, although the temperature of the internal boundary is variable over the boundary surface. The temperature of the external boundary, θ_e, is, by contrast, uniform and virtually constant.

Each junction of the pile (all of which are identical) occupies the same area, S, of the internal boundary wall, and there is left uncovered an area, S', between junctions.

The heat flux, Φ_1, from an area $(S + S')$ associated with a given thermocouple junction is only in part conducted away by the thermocouple wires. Let λ be the fraction of the total flux, Φ_1, conducted away in this way, and $(1-\lambda)$ the fraction lost by other means. λ is independent of the magnitude of Φ_1, and of the part of the boundary from which Φ_1 flows. Hence, we may write

$$\lambda\Phi_1 = C(\theta_i - \theta_e)$$

where C is the thermal conductivity of the thermocouple wires.

However, the e.m.f. developed in the thermocouple under consideration is

$$e_1 = \varepsilon(\theta_i - \theta_e)$$

where ε is the thermoelectric power of the thermocouple.

Combining these two relations, we obtain

$$e_1 = \frac{\varepsilon\lambda}{C}\Phi_1$$

Since all the couples are identical, the factor $\frac{\varepsilon\lambda}{C}$ is the same whichever couple is considered. The couples being connected in series, the total e.m.f. produced is the sum of the individual e.m.f's in each couple, e_1, e_2, e_3, etc. Hence

$$E = e_1 + e_2 + e_3 + \ldots\ldots = \frac{\varepsilon\lambda}{C}(\Phi_1 + \Phi_2 + \Phi_3 + \ldots.) =$$

$$= \frac{\varepsilon\lambda}{C}\Phi$$

where Φ is the *total heat flux* from the internal boundary.

Thus we see that if the thermoelectric pile is constructed as outlined above, the measured e.m.f., E, *is always proportional to the total flow of heat from the internal boundary,* irrespective of temperature fluctuations on the calorimeter wall.

CHAPTER 3

CONSTRUCTIONAL DETAILS OF THE MICROCALORIMETER

1. DESCRIPTION OF THE "MICROCALORIMETRIC ELEMENT"

The microcalorimetric apparatus is formed by the association of one or more pairs of "microcalorimetric elements", these being as far as possible identical, and maintained at the same temperature.

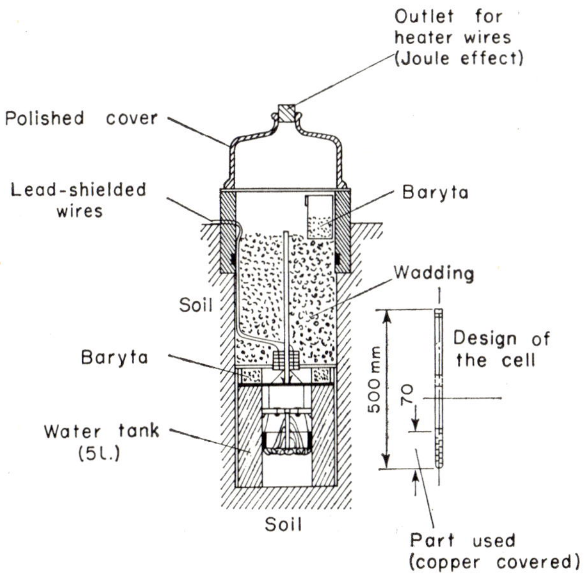

FIG. 3.1. Microcalorimeter of A. Tian, which was built into the ground.

The elements are constructed on the same principle as Tian's microcalorimeter, allowance being made for compensation by Peltier and Joule effects.

Each element has two thermoelectric piles, one serving to measure the heat flow between the two boundaries, and

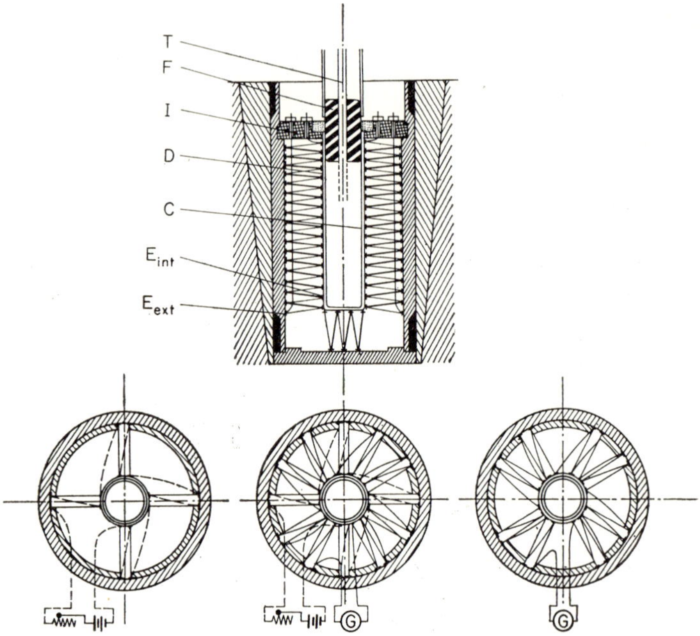

FIG. 3.2. Vertical and horizontal sections of a microcalorimetric element.

the other providing Peltier cooling. However, whereas Tian merely grouped the thermocouple junctions in a compact bundle round the internal wall, we have arranged them in a highly regular array, and both the number of thermocouples used, and the dimensions of the thermocouple wires, have been so chosen (by proper calculation) as to give optimum sensitivity of response and fidelity of performance. We have sought to obtain the minimal thermal inertia of the system to enable the apparatus to be used as far as possible as an "oscillograph".

Figure 3.2 shows vertical and horizontal sections of one of our most recent microcalorimetric elements.

The process under investigation is carried out in the cylindrical cell, *C*, which is either made of glass, with metallized exterior, or is made entirely of metal. The cell slides

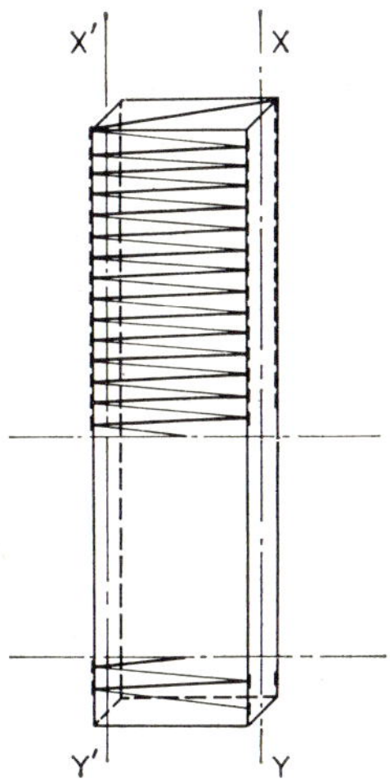

FIG. 3.3. Rectangular array containing two layers of thermocouples.

easily into a close-fitting cavity, *D*, made of silver. The cavity, *D*, is firmly fixed in the apparatus, but cell *C* can be removed as required, so that it can be charged outside the calorimeter in preparation for an experiment. The cavity, *D*, is quite thin, and is covered by a very thin insulating layer of mica (thickness 1/100 mm).

The surface of the cavity, *D*, constitutes the internal boundary, E_i. The external boundary, E_e, is the inside cylindrical wall of a hollow truncated cone, which is wedged into a cavity of the same shape in the calorimeter block, and firmly secured therein by a screwed ring. It is important to have very good thermal contact between the element and the block.

In our earlier apparatus, the thermoelectric piles were in the form of horizontal circular doughnuts: we now prefer to use vertical arrays of thermocouples, which are made by machine (Fig. 3.3).

2. The Experimental Zero of a Calorimeter containing only a Single Microcalorimetric Element

Let us consider a calorimeter of only a single element placed in a thermostat, and suppose that there is no thermal effect taking place in the calorimeter container.

The curve recorded by the movement of the galvanometer spot over a period of time under these conditions is referred to as the "*experimental zero*". It is to be distinguished from the "true zero", recorded by the galvanometer on open circuit. Differences between the position of the experimental zero and the true zero arise from continuous and irregular heat exchanges between the internal and external boundary due to imperfect thermostatting.

If the external boundary temperature is made to vary linearly with time, t, so that $\theta_e = \theta_e^\circ + \alpha t$, the temperature, θ_i, of the internal boundary follows suit, but in a delayed fashion as shown graphically in Fig. 3.4, becoming, after a certain time, linear and parallel to θ_e. The temperature of the internal boundary thus lags behind that of the external boundary by an amount $\alpha\tau$*, where τ is the *time constant* of the apparatus. The measurement of the magnitude of τ (which is one of the principal characteristics of the calorimeter) is not difficult, and is described further in Chapter 5.

Tian housed his microcalorimeter in the soil at the bottom of a deep cellar, where the temperature varied only

* This result is readily obtained by a simple calculation: see pp. 44–45 of *Microcalorimétrie*, Masson et Cie, Paris, 1956.

very slowly ($^1/_{100}$th degree per day). The experimental zero, however, rarely coincided with the true zero of the apparatus. It was assumed that the displacement was linear, and a straight line joining the recorded displacements of the experimental zero before and after a calorimetric experiment was adopted as the actual variation of the experimental zero. This procedure is an uncertain one, and the

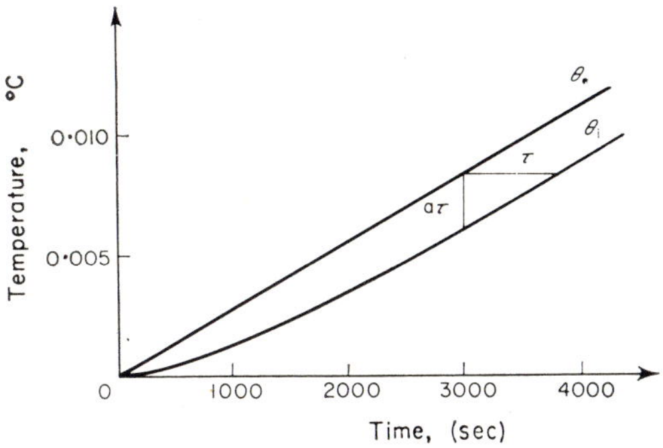

FIG. 3.4. The variation of θ_i relative to θ_e. when θ_e changes linearly with time.

usefulness of Tian's microcalorimeter was restricted to studies of fairly short duration—and certainly not to experiments lasting for more than a day.

These faults, responsible for the lack of fidelity of Tian s microcalorimeter, led us to adopt the "differential" arrangement of calorimetric elements, as our primary purpose has been to obtain a stable experimental zero.

3. The Differential Arrangement

As already pointed out, the differential arrangement employs two twin microelements with their thermoelectric piles connected together in "opposition", the leads from the external boundaries being connected together, whilst those from the internal boundaries are connected to the galvanometer. The thermal process under investigation is carried out in one of the elements (which we call the *laboratory* element), the second element serving as the *reference* element. Compensation by the Peltier effect is made in the laboratory element, so as to annul the heat developed by the process occurring therein. If compensation is total, neither of the twin elements liberates any heat. In this circumstance, the experimental zero is quite stable, as shown below:

The temperature of the external boundaries being the same, θ_e, for both elements at any given time, the e.m.f's developed in the piles are:

$$e_1 = n\varepsilon(\theta_{i1} - \theta_e) \quad \text{for the first element}$$
$$e_2 = n\varepsilon(\theta_{i2} - \theta_e) \quad \text{for the second element}$$

If the temperature of the block containing the twin elements begins to vary at a certain time by an amount $\Delta\theta_e$, the internal boundaries are affected identically by an amount $\tau\Delta\theta_e$ (the time constants τ of the elements being essentially the same). Thus, the modified e.m.f's become:

$$e_1 = n\varepsilon[(\theta_{i1} + \tau\Delta\theta_e) - (\theta_e + \Delta\theta_e)]$$
$$e_2 = n\varepsilon[(\theta_{i2} + \tau\Delta\theta_e) - (\theta_e + \Delta\theta_e)]$$

The total e.m.f. for the system in opposition is

$$e = e_1 - e_2 = n\varepsilon(\theta_{i1} - \theta_{i2}) \qquad (13)$$

which is independent of *variations* in the temperature of the external boundaries.

4. Temperature Constancy of the External Boundary

The differential arrangement of twin elements in the same thermostatted block resolves the major difficulty in the achievement of fidelity of performance, by giving an experimental zero which is perfectly stable over long time-periods.

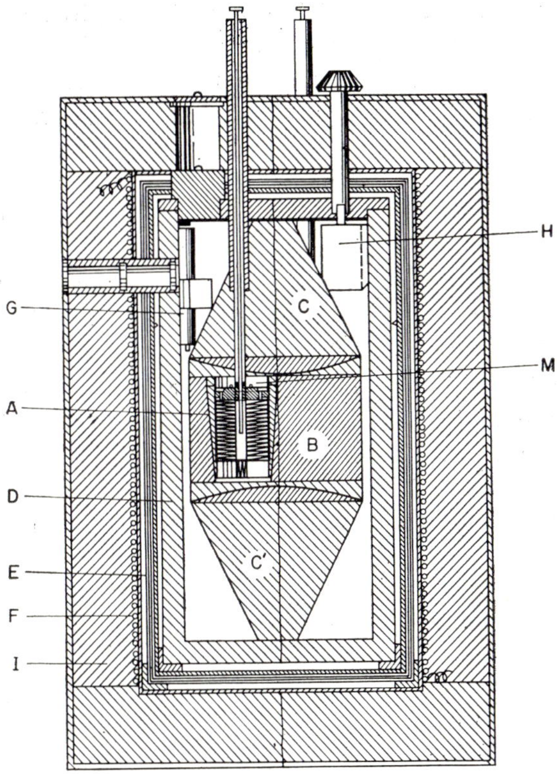

FIG. 3.5. Vertical cross-section of a recent model of Calvet's apparatus.

A = Conical socket cut into the block B.
B = The metallic block.
C,C' = Metallic cones designed to give equipartition of the thermal perturbations.
D = Thick metal cylinder.
E = Thermostat consisting of several concentric canisters.
F = Electrical heater controlled by the thermoregulator.
G = Galvanometer.
H = Switch
I = Thermal insulation.
M = Microcalorimetric element (placed inside its cavity A).

It demands only that the temperatures of the two external boundaries should be *identical at any given instant*; to achieve this condition is a much easier task than that of keeping them both at one *fixed* temperature.

We have arranged the apparatus as shown diagrammatically in Fig. 3.5.

The calorimetric elements fit into the conical sockets, A, cut in a cylindrical metallic block of silver or aluminium, B. An even number (usually 2 or 4) of elements is used, these being housed symmetrically in the block. The block itself sits between the bases of two truncated cones, *C*, *C′*, centrally placed within a thick metal cylinder, *D*, surrounding the block. This cylinder is in turn enclosed in a carefully controlled thermostat built up of several concentric metal canisters.

The purpose of this arrangement is to transform lateral thermal perturbations not confined to the symmetry plane of the twin elements into vertical pertubations equally divided between the twin elements. A lateral perturbation is, in the first place, spread out by the multiple jackets of the thermostat, diffused in the thick outer cylinder wall, and conducted away from the top of the metal cones symmetrically towards the twin elements. Thus, the cones act as collimators to transform the heat flux from the cone tops into a parallel flux through the block. We have even been able successfully to use thermal lenses, made of poorly conducting steel, which achieve a practically perfect equipartition of the thermal perturbations*. The disposition of the thermal lenses is shown in Fig. 3.5.

As long as the twin elements are identical in construction, and symmetrically disposed with respect to the central axis of the cylindrical block, it is not necessary to maintain complete temperature uniformity within the block. It suf-

* See *Microcalorimétrie*, by E. Calvet and H. Prat, p. 56 (Masson et Cie, Paris, 1956).

fices to ensure that the temperature distribution is symmetrical at corresponding points on the two elements.

It is by using the cones as equi-distributors that we have been able to achieve an apparatus of high fidelity, and furthermore, to develop it for calorimetric studies at high temperatures.

CHAPTER 4

THEORY AND MANIPULATION OF THE APPARATUS

1. The Case of Slow Processes

In the case of processes which cause only slow variations in the temperature of the laboratory element, the simplified theory applied by Tian to his single-element microcalorimeter can be used in good approximation.

In the more general case, we will show that Tian's simple equation is still a good approximation, subject to certain corrections being made to the directly recorded curves.

(a) Tian's equation

The equation relates to the ideal case in which the calorimeter contents, and all points on the internal boundary, are at the same temperature at a given time. This is equivalent to assuming that the element and its contents have an infinite thermal conductivity.

In this ideal case, let W be the calorific power developed in the cell at time, t. W is in part lost from the cell as a flow of heat, Φ, which is proportional to the temperature difference $(\theta_i - \theta_e) = \theta$, between the internal and external boundaries, i.e.

$$\Phi = p(\theta_i - \theta_e) = p\theta \qquad (14)$$

The remaining part of W causes an increase, $d\theta$, in the internal temperature (assuming the temperature, θ_e, is constant). If the thermal capacity of the cell contents is μ, the internal heating consumes power at the rate $\mu \dfrac{d\theta}{dt}$. Hence, it follows

$$W = p\theta + \mu \frac{d\theta}{dt} \tag{15}$$

The galvanometer in the detector pile circuit shows deviations, Δ, (which are directly recorded) proportional in magnitude to θ,

$$\Delta = g\theta \tag{16}$$

presuming that the mechanical inertia of the galvanometer suspension, and the self-induction of the circuit can be neglected. On combining (16) and (15), we obtain Tian's equation

$$W = \frac{p}{g}\Delta + \frac{\mu}{g}\frac{d\Delta}{dt} \tag{17}$$

(b) Thermal disequilibrium and the effective heat capacity

Internal thermal disequilibrium – To assume complete temperature uniformity within the contents and on the walls of the internal cell, is clearly unacceptable in practice. Actually, considerable temperature gradients would exist in a cell containing a heat source surrounded by a substance of poor thermal conductivity. In general, we have to consider the temperature distribution to be heterogeneous, and a state of *internal thermal disequilibrium* to exist.

Nevertheless, when the thermal process occurring in the cell is slow, it is still permissible to make use of Tian's equation, provided that the true heat capacity is replaced by an *effective heat capacity.*

When a heat source of constant power, W, contained inside the cell is cut off, the total heat liberated as the cell cools down to the external temperature, θ_e, exceeds the quantity $C\theta$ (where C is the true heat capacity, and θ the temperature difference between the internal and external boundaries) by an amount Q_i, which owes its origin to the temperature gradients existing between cell contents and the boundary. One can regard Q_i as the quantity of heat that would be released should the thermal conductivity of the internal cell and its contents suddenly become infinite. We shall refer to Q_i as the *heat of internal thermal disequilibrium,* and it can be shown* (and experimentally verified) that Q_i is proportional to W, so that

$$Q_i = D_i W \qquad (18)$$

D_i, which has the dimensions of time, is known as the *coefficient of internal thermal disequilibrium.*

External thermal disequilibrium — It is also necessary to allow for a heat of external thermal disequilibrium, Q_e, which exists when heat accumulates in the space between the two boundaries. Q_e is also proportional to W,

$$Q_e = D_e W \qquad (19)$$

where D_e is the coefficient of external thermal disequilibrium.

In fact, in our most recent apparatus, we have succeeded in suppressing the external thermal disequilibrium by using short, bare wires, very regularly disposed round the cell, and by markedly reducing the thermal inertia of the internal thermoelectric junctions**. The total heat of thermal disequilibrium is, however, generally to be written as

$$Q = Q_i + Q_e = D_i W + D_e W = (D_i + D_e) W = DW \qquad (20)$$

D being the coefficient of *total* thermal disequilibrium.

It can be shown*** that the effective heat capacity, μ, is related to the true heat capacity, C, by the formula $\mu = C + Dp$, D being the coefficient of total thermal dis-

* See *Microcalorimétrie,* p. 74.
** See *Microcalorimétrie,* p. 75 onwards.
*** See *Microcalorimétrie,* p. 78.

equilibrium, and p = thermal conductivity of the barrier separating the internal and external boundaries. It is the effective heat capacity, μ, which must be used in Tian's equation when it is applied to cases where there is internal thermal disequilibrium.

(c) Different methods of using the apparatus

(i) **Thermal compensation within the laboratory cell** — We have already seen, in the differential arrangement, that compensation must be made in the laboratory-element, the reference-element remaining throughout at the temperature of the block. If the calorific power produced by compensation (usually, the Peltier effect) is W', Tian's equation takes the form:

$$W = W' + \frac{p}{g} \cdot \Delta + \frac{\mu}{g} \frac{d\Delta}{dt} \tag{21}$$

(ii) **Use of the apparatus to measure heat output between times, t and t' (use as an integrator)** — The quantity of heat $Q_t^{t'}$ produced in the time interval $(t' - t)$ is obtained from equation (21) by integration:

$$Q_t^{t'} = \int_t^{t'} W dt = \int_t^{t'} W' dt + \frac{p}{g} \int_t^{t'} \Delta dt + \frac{\mu}{g} \int_t^{t'} d\Delta \tag{22}$$

The term $\int_t^{t'} W'dt$ is the heat compensated, which we shall represent by C.

As already stated, total compensation is not recommended, and it is preferable to make only partial compensation of the process occurring in the laboratory-element. The Peltier compensation currents, $i_1, i_2, i_3, \ldots, i_j$, are applied for measured times, $t_1, t_2, t_3, \ldots, t_j$, and the total compensation heat obtained by summing the individual contributions, $W'_1 t_1$, $W'_2 t_2$, ...

$$C = \sum_{i=1}^{i=j} W'_i t_i \tag{23}$$

The compensated heat should account for 90% of the total heat to be measured, and is capable of very precise measurement (the Peltier currents i_1, i_2, etc., being accurately determined by potentiometric methods). If the non-compensated heat is evaluated within error limits of 2% or less, the overall error attached to the determination of the *total* heat is then not more than 2 parts in a thousand.

The integral, $\int_t^{t'} \Delta dt$, in the second term of equation (22), is equal to the area, A, under the recorded curve and the time-axis over the interval $(t' - t)$. An area situated above or below the time-axis is considered positive or negative, corresponding to a heating or cooling effect.

The third term in (22) contains the integral $\int_t^{t'} d\Delta$, which is the difference $\Delta' - \Delta$ in the galvanometer deflections at times t' and t. Hence, we may rewrite (22) as:

$$Q_t^{t'} = C + \frac{p}{g} A + \frac{\mu}{g} (\Delta' - \Delta) \tag{24}$$

For precise measurements, in which only a small fraction of the total heat is not compensated, the last two terms of equation (24) can be regarded as correction terms.

The term $\frac{p}{g} A$, which we refer to as the "*area correction*", can be determined with precision in most cases (see Chapter 5). The term $\frac{\mu}{g} (\Delta' - \Delta)$ is more difficult to determine accurately: this we refer to as the "*heat capacity correction*". It is, however, possible to make this correction small, and even zero, if t, t' are chosen so that $\Delta = \Delta'$.

Hence it is not necessary to attempt total compensation in order to obtain a high precision (> 1/1000). In fact, it is possible to obtain a precision better than 1% without any compensation at all, provided that the experiment is not of short duration, and the heat capacity correction is small. The apparatus is particularly well-suited for the study of slow thermal processes.

(iii) **The use of the apparatus as an oscillograph, for thermokinetic studies** — From equation (21), we may write:

$$W - W' = \frac{p}{g}\Delta + \frac{\mu}{g}\frac{d\Delta}{dt} = \frac{p}{g}\left(\Delta + \frac{\mu}{p}\frac{d\Delta}{dt}\right) =$$

$$= \frac{p}{g}\left(\Delta + \tau\frac{d\Delta}{dt}\right) \qquad (25)$$

where $\tau = \frac{\mu}{p}$, is the *time-constant* of the apparatus. The method of evaluating τ will be described later.

The direct record of the galvanometer deflection gives Δ as a function of time. The thermokinetic curve, however, relates W to time. To transform one curve into the other, we add to each point Δ a correction term which is the product of the time-constant τ and the slope of the curve at that point. Thus we obtain a new curve (subject to the scale-factor $\frac{p}{g}$) which gives the non-compensated calorific power, $W - W'$, as a function of time. The thermokine tic curve, $W = f(t)$, is then obtained by adding the compensation W', separately measured.

In the case of very slow thermal phenomena, for which the correction $\tau\frac{d\Delta}{dt}$ is very small relative to Δ, the directly recorded curve can be accepted as a satisfactory representation of the thermokinetics of the process under investigation.

It should be pointed out that the suitability of the apparatus for use as an *oscillograph* (i.e. giving a direct recording of the thermokinetics) depends on the smallness of the time-constant, τ; this condition is best met by a small effective heat capacity, μ, and a large coefficient of thermal leakage, p, so that it is best to employ small quantities of material, and to reduce the sensitivity of the apparatus.

By contrast, when the apparatus is being used as an *integrator*, the thermal leakage should be small, and the heat capacity large—i.e. one needs a large time-constant.

2. The Case of Relatively Rapid Thermal Processes

Tian's equation is not applicable directly to the study of rapid processes which are accompanied by thermal changes. We have therefore sought a method of transforming the directly recorded curves so that they become suitable for treatment by Tian's simple equation.

(a) Analysis of the recorded curves

(i) **Ballistic phenomena** — Suppose that at time $t = 0$, a constant thermal power W is generated in the cell, and continued for a heating period, u, before being cut off. The shape of the recorded curve, $\theta = \Phi(t)$, depends upon the duration of the heating period. When u is relatively long, the recorded curve takes the form of the last curve shown in Fig. 4.1. The rising part of this curve is approximately exponential, and follows the equation

$$\theta = \theta_0 \ (1 - e^{-\omega t}) \tag{26a}$$

where θ_0 is the maximum height reached during the linear portion of the curve. The falling part of the curve is also exponential, obeying the equation

$$\theta = \theta_0 \, e^{-\omega(t-u)} \tag{26b}$$

The coefficient, ω, is a constant of the calorimeter, the inverse, $1/\omega$, being the time-constant, τ. In our present apparatus τ is of the order of 8 min; the maximum height, θ_0, is reached after a heating period, $u = 4\tau$.

When the thermal power is applied for relatively short times of the order τ, or less than τ, one obtains the remaining curves reproduced in Fig. 4.1. It is noticeable that short heating periods give curves with sharp pointed maxima. We have shown that the heights of these maxima are proportional to the total heat developed, $Q = \int_0^u W\, du$. The figure shows the observed curves, θ, and the actual heat outputs, W (dotted rectangles).

In most cases it is possible to analyse the "shock curves" in terms of two exponential functions. The breakdown of

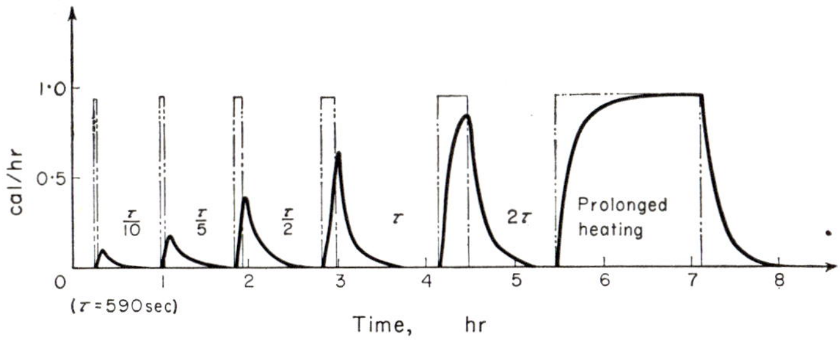

FIG. 4.1. Recordings given by applying constant thermal power, W, for heating periods $\tau/10$, $\tau/5$, $\tau/2$, τ, 2τ, and for a long period.

a typical ballistic curve, $OABC$, into two exponentials, DBC, and EF, is shown diagrammatically in Fig. 4.2.

The curve, $OABC$, shows the effect of a thermal shock, W produced at time $t = 0$ for a very short heating period, du. The curve follows the equation

$$\theta = Wa(e^{-\omega_1 t} - e^{-\omega_2 t})\, du \tag{27}$$

The descending part, BC, is an exponential curve of time-constant $\tau_1 = 1/\omega_1$; on extrapolation to zero time, the curve reaches point D, ordinate $= Wadu$. By subtracting the extrapolated curve, BD, from the recorded curve, OAB, the dotted curve, EF, is obtained. This is an exponential curve, with time-constant $\tau_2 = 1/\omega_2$, and with the ordinate $OE = OD$.

A more precise theory of thermal shock shows that the "ballistic curve" can be expressed by a series of exponentials of the form

$$\theta = \Delta Q(a_1 e^{-\omega_1 t} + a_2 e^{-\omega_2 t} + \ldots\ldots) \tag{28}$$

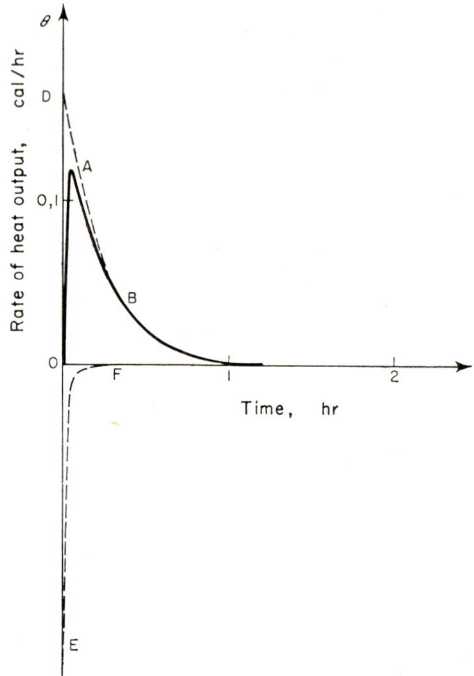

FIG. 4.2. Breakdown of a recorded ballistic curve into two exponentials.

where ΔQ is the heat generated by the shock. In general, only the first two terms of the series are needed to give a satisfactory account of our calorimeters, since the time-constants, ω, diminish rapidly on ascending the series.

(ii) **General case** — Let us suppose that an infinitesimal thermal impulse, $W du$, produced in the cell at time u, results in an increase in temperature, $d\theta$, at time t. The general theory enables us to write:

$$d\theta = W du(a_1 e^{-\omega_1(t-u)} + a_2 e^{-\omega_2(t-u)} + \ldots .) \qquad (29)$$

For a longer and complex heating in the cell, W is a function of the time of heating, u. We distinguish the time, u, from the time of observation, t, of the temperature, θ. The effect of applying heating power $W = f(u)$ can be regarded as the integration of a succession of thermal pulses $W du$, so that to obtain $\theta = \Phi(t)$, we have

$$\theta = a_1 \int_0^t W e^{-\omega_1(t-u)}\, du + a_2 \int_0^t W e^{-\omega_2(t-u)}\, du + \ldots . \qquad (30)$$

As already stated, only the first two terms of this series are normally required. In this event $a_1 = a_2$, as follows from (28) and the condition that $\theta = 0$ at $t = 0$. Accordingly, we can replace (30) by equation (31):

$$\theta = a \int_0^t W e^{-\omega_1(t-u)}\, du - a \int_0^t W e^{-\omega_2(t-u)}\, du \qquad (31)$$

Writing

$$\theta_1 = a \int_0^t W e^{-\omega_1(t-u)}\, du \qquad (32)$$

and

$$\theta_2 = a \int_0^t W e^{-\omega_2(t-u)}\, du \qquad (33)$$

it follows from (31), and by differentiation of (31), that

$$\theta = \theta_1 - \theta_2 \qquad (34)$$

$$\frac{d\theta}{dt} = -\omega_1 \theta_1 + \omega_2 \theta_2 \qquad (35)$$

Hence θ_1 can be expressed as a function of θ,

$$\left(\frac{\omega_2 - \omega_1}{\omega_2}\right)\theta_1 = \theta + \frac{1}{\omega_2}\frac{d\theta}{dt} \qquad (36)$$

The curve of θ_1 can be drawn from the curve of θ in the following way:

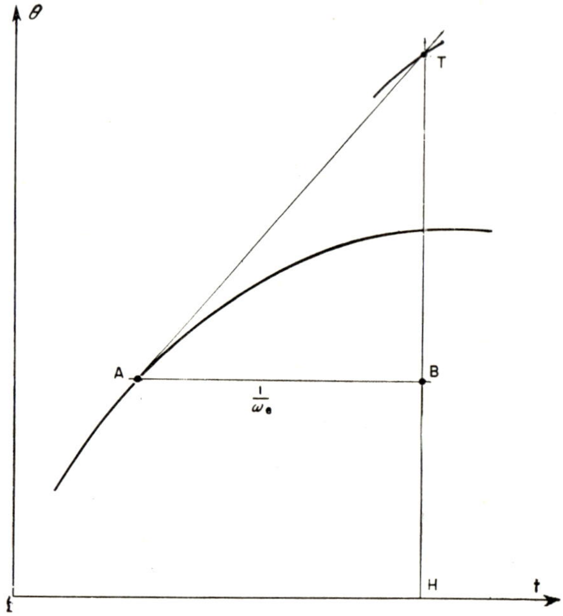

FIG. 4.3

Let A be a point on the curve $\theta = \Phi(t)$. Draw AB parallel to $\overline{Ot}$, such that $AB = \frac{1}{\omega_2}$. Extend the tangent at A to AT, HBT being perpendicular to AB.

The vector $\overline{HB}$ represents θ, and the vector $\overline{BT}$ represents $\frac{1}{\omega_2} \frac{d\theta}{dt}$. Hence the ordinate, T, has the value $\left(\theta + \frac{1}{\omega_2} \frac{d\theta}{dt}\right)$, which, from eqn. (36), equals $\frac{\omega_2 - \omega_1}{\omega_2} \theta_1$.

Hence the curve through the points T represents the function $\theta_1 = \Phi_1(t)$, bearing in mind the multiplying factor $\left(\frac{\omega_2 - \omega_1}{\omega_2}\right)$.

On differentiating equation (32), we have:

$$\frac{d\theta_1}{dt} = -\omega_1 a e^{-\omega_1 t} \int_0^t W e^{\omega_1 u} du + aW_{(u=t)} =$$

$$= -\omega_1 \theta_1 + aW \qquad (37)$$

or, after rearranging,

$$\frac{a}{\omega_1} W = \theta_1 + \frac{1}{\omega_1} \frac{d\theta_1}{dt} \qquad (38)$$

This is Tian's equation, and is of the same form as equation (36): hence the operations made on the curve of θ to produce θ_1 now carried out on the curve of θ_1 produce the curve of W, multiplied by the factor $\frac{a}{\omega_1} \cdot \frac{(\omega_1 - \omega_2)}{\omega_2}$.

The transformation of the directly observed curves of θ into curves of W is only necessary for processes in which the heat output varies sharply with time: slow thermal outputs can be dealt with by direct application of Tian's equation.

(b) A micro bomb calorimeter

The process of combustion of an organic substance in a metal bomb filled with oxygen under pressure produces a very rapid heat output, and the bomb calorimeters normally used for measuring heats of combustion are "macro" calorimeters, of *ca.* 5000 cal/°C heat capacity. Several grams of pure substance are required in order to measure a heat of combustion accurately in a normal sized bomb calorimeter: there is, therefore, something to be said for the development of a "micro" bomb calorimeter, requiring much smaller amounts of substance, which would have special value for the study of compounds difficult to obtain in the pure state in quantity.

However, the Calvet microcalorimeter, as normally used, is too sensitive an instrument for combustion studies. For instance, the combustion of 0.01 g benzoic acid liberates 63 calories: the microcalorimeter responds with a scale deflection of 1 mm (on a scale 1 m from the galvanometer)

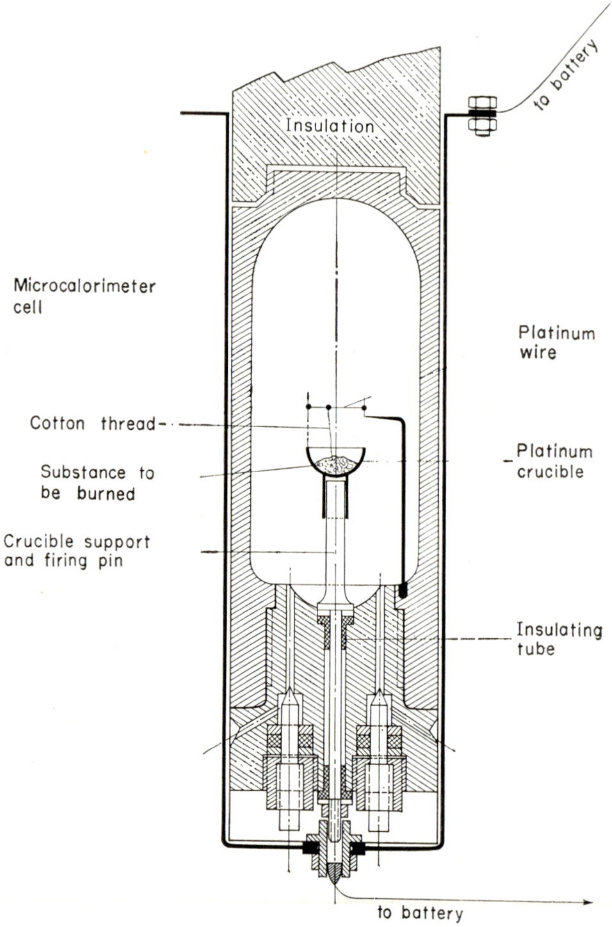

FIG. 4.4.

to an output of only 0.001 cal. per hour, so that to record a satisfactory thermogram of the combustion of benzoic acid, only a milligram or so would be required. But the accuracy of weighing such a small quantity falls well below the precision of 1 part in 10^4 presently sought in modern "macro" bomb calorimetric studies.

The sensitivity of the microcalorimeter can of course be reduced by using a less sensitive galvanometer or by replacing the galvanometer with a commercial recording potentiometer. Using the latter, we have recorded satisfactory thermograms given by burning benzoic acid samples weighing *ca.* 0.01 g.

The early experiments were made with cylindrical "semi-micro" bombs, fitted into a laboratory cell of 35 mm diameter, 80 mm height. Several micro bombs of different sizes were tried, but the very small bombs proved unsatisfactory since the flame reached the bomb walls resulting in incomplete combustion and carbonaceous deposits. This difficulty can, however, be largely overcome by coating the bomb walls with a thermal insulator, such as alumina or even mica. Semi-micro bombs proved to be generally the most satisfactory, mainly because they support the combustion of larger amounts of substance than bombs of "micro" dimensions.

A cross-sectional diagram of a satisfactory experimental micro bomb is shown in Fig. 4.4.

(c) Design of a micro bomb combustion calorimeter

The bomb was designed to fire in the "inverted" position, this giving protection from the direct heat of the flame to the delicate valves fitted into the base of the bomb. The combustion sample was contained in a platinum crucible, centrally supported in the bomb above the inlet and exit valves. The sample was fired electrically in the usual way, using a small piece of cotton fuse.

In each combustion experiment, two identical micro-bombs were used, one in the laboratory cell, the other in the reference-cell of the differential calorimeter. Both bombs were fitted identically with platinum heater wire, and cotton fuse, but only the laboratory cell contained the combustion sample. Both bombs were fired simultaneously by the same electric current. This differential firing arrangement has the advantage of eliminating the correction normally required for the ignition energy.

The coating of the inside walls of the bomb and crucible with a thermal insulator not only prevents incomplete combustion of the sample, but also *reduces the thermal diffusivity* of the micro bomb. This latter effect is of considerable importance, since, by interposing a thermal insulating layer between the bomb and surrounding thermoelectric pile, *the ballistic curves are damped and become representable by a sum of only two exponential terms.* Hence the task of analysing the recorded thermograms is very much simplified. In fact, when the duration of thermal shock is very short, as is the case with a combustion experiment, the maximum deviation recorded on a thermogram is proportional to the heat liberated. This simple method of measurement is capable of a precision of *ca.* 1 part in 200.

Although falling well below the degree of precision attainable with macro-scale bomb calorimeters, that attained so far with the micro bomb is encouraging, and one can hope that further improvements in design and operation will render the instrument of increasing value in this important field of thermochemical study.

CHAPTER 5

CALIBRATION OF THE MICROCALORIMETER EXPERIMENTAL DETERMINATION OF THE CONSTANTS OF THE APPARATUS

We shall only consider curves that have been adjusted to conform with Tian's equation.

1. Calibration of the Galvanometer Deviation: The Measurement of p/g

When a constant calorific power W is generated in the laboratory-cell without compensation, there results a deviation Δ of the galvanometer, which becomes constant $\left(\text{i. e.} \frac{d\Delta}{dt} = 0\right)$ once the steady state is established (see Fig. 5.1). In this condition, Tian's equation (17) reduces to the form: $W = \frac{p}{g}\Delta$.

The coefficient p/g is the principal constant of the apparatus; a knowledge of its value is all that is needed when the apparatus is used without compensation, bearing in mind that the heat capacity correction can either be ignored (slow processes) or rendered zero (identity of initial and final deviations).

To measure p/g, one records first of all the trace of the experimental zero, OA (Fig. 5.1), then the power production is begun at time A, either by means of the Joule effect, or by introducing a needle of radium centrally into the cell. The curve $ABCD$ is traced out, becoming parallel to the time axis in CD. The constant deviation, Δ, is the height Aa in the figure.

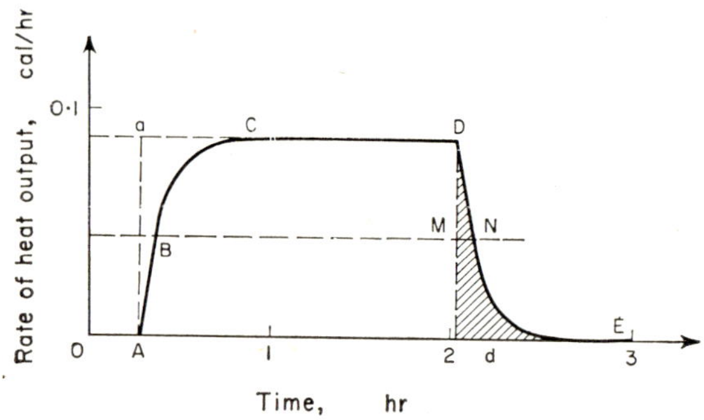

FIG. 5.1. Measurement of the ratio $\frac{p}{g} = \frac{W}{\Delta}$

The Joule effect provides a very convenient source of the calorific power, W, because it is easy to regulate. However it has some disadvantages, namely:

(a) The leads to the electric heater in the laboratory-cell allow thermal leakage from the cell which is not negligible.

(b) The precise measurement of the Joule effect needs expensive electrical equipment.

The calibration is made more easily by using radioactive standards (of long half-life period) as the source of constant calorific power, W, since their heat output is *independent of temperature*.

We use a platinum tube containing from 0.5 to 1 mg of radium as the standard generator of calorific power. The platinum tube totally absorbs the emitted α- and β-radiation, as well as the soft γ-rays. Figure 5.2 shows the simplicity of this device: the radioactive source, R, is suspended from a glass-rod terminating in a ground-glass stopper, B; the thin glass tube, T, passing through the neck of the laboratory-cell, houses the source, R, when it is introduced into the cell.

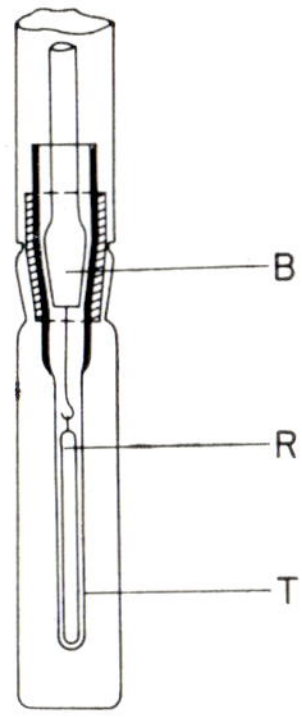

FIG. 5.2. The use of a radioactive source for calibration purposes.

2. The Measurement of μ/g

We will now consider the part DE of Fig. 5.1, which represents the return of the galvanometer deflection to the experimental zero after cutting off the power, W, in the laboratory-cell.

Since no heat is produced in the cell during the period DE, nor is any compensation made, the equation (24) becomes:

$$0 = \frac{p}{g} A - \frac{\mu}{g} \Delta_0$$

A being the shaded area, DdE, and Δ_0 the deviation, Dd, at the instant of suppression of the power, W. Hence,

$$\frac{\mu}{g} = \frac{p}{g} \frac{A}{\Delta_0} \tag{39}$$

in which p/g is already known following the calibration of the galvanometer deviation.

3. The Measurement of the Peltier Coefficient, P

As shown in Chapter 2, the thermal power produced by the Peltier compensation effect is given by:

$$W = Pi\left(1 - \frac{i}{I}\right) \tag{40}$$

P is measured from the current i required in order completely to compensate a constant heat power, W, liberated in the laboratory cell. The value of I in equation (40) is separately determined, this being the Peltier current for which the Peltier cooling at the "cold" junctions is annulled by the self-heating due to the flow of the current (p.14).

4. Measurement of the Coefficient of Thermal Disequilibrium, D

Definitions of the "heat of thermal disequilibrium" and of the "coefficient of thermal disequilibrium" have already been given in Chapter 4. They may be measured thus:

A constant calorific power, W, is produced in the laboratory-cell by the Joule effect, and is totally compensated by the Peltier effect: this is not difficult to do if a variable rheostat is included in the Peltier current circuit, and adjusted to maintain the galvanometer deflection at the experimental zero. If now both Peltier and Joule currents are simultaneously cut off, the immediate galvanometer response indicates a heating effect (see Fig. 5.3). The area under the recorded curve measures the heat of thermal disequilibrium, Q, from which the coefficient of thermal disequilibrium may be derived: $D = \frac{Q}{W}$.

During the compensation, the internal and external temperatures are kept, effectively, at the same temperature: but, when the Joule and Peltier effects are simultaneously cut, the *accumulated* heat in the cell, due to thermal disequilibrium, begins to escape. Hence the *heating* effect, *BCD*, in Fig. 5.3. If now, on the return to the experimental zero,

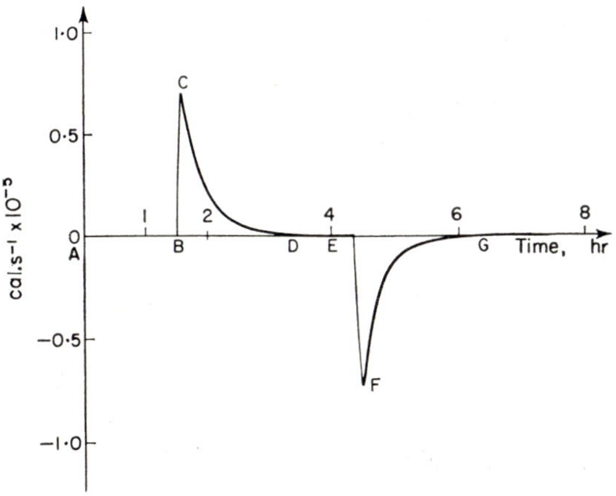

FIG. 5.3. Recording of the heat of thermal disequilibrium. *AB* is the experimental zero, recorded whilst a power of 57×10^{-6} cal.sec.$^{-1}$ was developed and totally compensated in the cell. At *B*, both Joule and Peltier currents were cut off. *BCD* is the resulting curve, corresponding to a heat evolution. At *E*, the Joule and Peltier currents were switched on once more. *EFG* is the resulting curve, corresponding to heat absorption by the cell contents.

the *same* Joule and Peltier currents are simultaneously switched on, the galvanometer trace shows a *cooling* effect, identical in shape and area to *BCD*, but opposite in sign, *EFG*, of Fig. 5.3. The two curves *BCD* and *EFG* are exactly superposable if the external disequilibrium due to the flow of the Peltier current is nil.

4

CHAPTER 6

SENSITIVITY, REPRODUCIBILITY, AND PRECISION OF THE MICROCALORIMETER

1. The Sensitivity of the Microcalorimeter

The sensitivity of the apparatus is most easily studied in the case of very slow thermal phenomena, when the galvanometer deviation, Δ (in the absence of compensation) is directly proportional to the power, W, produced in the cell. In this circumstance, Tian's equation (17) reduces to a single term

$$W = \frac{p}{g}\Delta \tag{41}$$

We shall restrict ourselves in discussion to this particular case, and draw a number of conclusions enabling us to define the conditions necessary for maximum sensitivity, relating to the nature and the number of thermocouples required, the length and diameter of the wires, and the thermal isolation of the cell*.

(a) Definition of the sensitivity of the microcalorimeter

The galvanometer deflection, Δ, is proportional to the value I of the current flowing in the galvanometer circuit,

$$\Delta = sI \tag{42}$$

* The calculation of the sensitivity of the apparatus was made in collaboration with M. Persoz, of the Office National des Recherches Aéronautiques. For details, see *Microcalorimétrie*.

which, substituted into equation (41), gives:

$$W = \frac{p}{g} sI \tag{43}$$

The sensitivity, σ, of the microcalorimeter is defined by

$$\sigma = \frac{I}{W} \tag{44}$$

We shall now calculate I and W for the particular case of a differential microcalorimeter of two elements. The e.m.f. of the reference element is zero, but the electrical resistances of the two elements add together.

The calculation of I, for a temperature difference, θ, between the internal and external boundaries, depends on the thermoelectric power of the thermocouples and the electrical characteristics of the circuit. The calculation of W as a function of θ depends on the thermal characteristics of the medium separating the two boundaries. We have also to allow for the Peltier effect due to the current, I, in the thermopile.

We shall consider heat loss by conduction only, and neglect radiation and convection heat losses.

The following notation is adopted:

S and S' are cross-sections of the wires constituting the couples.

l is the common length of these wires.

n is the number of couples constituting the thermopile of a microcalorimetric element (for maximum sensitivity, the two piles are joined in series).

c and c', r and r' are thermal conductivities, and electrical resistivities of the materials used for the thermocouple wires.

p is the heat loss from the calorimeter per second, per degree temperature difference between the boundaries.

p_c is the heat loss by conduction by the thermocouple wires, and p_0 the heat loss by other causes, $(p = p_c + p_0)$.

ϱ is the resistance of the galvanometer and the connecting wires.

E_0 is the thermoelectric power of a couple.

π is the heat absorbed by the Peltier effect per second in each junction when traversed by a current of 1 ampere.

$$\text{Let } R = \frac{r}{S} + \frac{r'}{S'}, \quad C = cS + c'S'$$

Calculation of I — The total e.m.f. of the detector pile is

$$E = nE_0\,\theta$$

The resistance of the circuit is

$$\varrho + 2nlR$$

$$\therefore\; I = \frac{nE_0\theta}{\varrho + 2nlR} \tag{45}$$

Calculation of W — When the steady state is established, the power W produced in the laboratory-element is in very small part compensated by the Peltier effect due to the current, I, to the extent $n\pi I$. A similar Peltier compensation applies to the reference-element

$$\therefore\; W = 2n\pi I + p\theta \tag{46}$$

$$\text{where } p = p_0 + p_c = p_0 + \frac{nC}{l} \tag{47}$$

The sensitivity expression — By combining equations (45), (46), (47) and (44), we find

$$\sigma = \frac{E_0}{2n\pi E_0 + \dfrac{\varrho p_0}{n} + \dfrac{\varrho C}{l} + 2lRp_0 + 2nRC} \tag{48}$$

The term $2n\pi E_0$ in the denominator is negligible in comparison with the other terms. The maximum sensitivity is obtained when the denominator is a minimum, and the conditions for maximum sensitivity result by setting the differentials of the denominator with respect to n, l, S, and S' to zero. If the common length, l, of the wires is fixed, the following maximum sensitivity conditions apply:

$$n = \frac{\sqrt{\frac{1}{2}\varrho \cdot p_0}}{\sqrt{cr} + \sqrt{c'r'}} \tag{49}$$

$$S = l\sqrt{\frac{2p_0 r}{\varrho \cdot c}} \tag{50}$$

$$S' = l\sqrt{\frac{2p_0 r'}{\varrho \cdot c'}} \tag{51}$$

Hence, we see that the optimum number of couples does not depend on the length of the wires, whilst the optimum cross-sections are proportional to the length.

Substituting the above relations into equation (48), the maximum sensitivity obtainable is expressed by

$$\sigma = \frac{E_0}{2n\pi E_0 + 4\sqrt{2}\sqrt{\varrho p_0}(\sqrt{cr} + \sqrt{c'r'})} \tag{52}$$

The examples we have cited in *Microcalorimétrie* (Masson et Cie, Paris) show that it is not difficult to obtain a sensitivity such that a few tenths of a microwatt cause a deviation of 1 mm. on a scale placed at 1 meter from the galvanometer spot.

(b) Choice and nature of the thermocouple wires

Neglecting the term $2n\pi E_0$ in (48), the sensitivity increases by choosing metals for which the magnitude of the quantity

$$\frac{E_0}{\sqrt{cr} + \sqrt{c'r'}} \tag{53}$$

is largest.

The products cr do not vary much from one metal to another (Wiedemann–Franz law): hence, it would seem that the important factor is the thermoelectric power E_0. However, substances of high thermoelectric power (tellurium, germanium, silicon, etc.) generally have high specific resistance, so that wires of thick cross-section would be needed. This is disadvantageous, since it leads to a large coefficient of external thermal disequilibrium. We have therefore made use of alloys, which, because they deviate from the Wiedemann–Franz law, increase the magnitude defined by (53).

2. The Reproducibility of the Apparatus

(a) Definition of the fidelity of the microcalorimetric apparatus

The high sensitivity of our microcalorimeter is useful only in so far as it is matched by a high reproducibility of performance—i.e. the recorded galvanometer deviations must be the same for the same power production in the calorimeter under identical conditions of operation. There are three different ways of testing the fidelity of the apparatus:

(i) *The power produced is zero:* in this event, the recorded curve is the *experimental zero,* which should be a perfectly straight line parallel to the time-axis. The record of the experimental zero, for different sensitivity conditions, is an excellent test of the fidelity of the apparatus. In the earliest microcalorimeters of a single element, the fluctuations in the experimental zero were too great to allow investigations lasting for more than a few hours. Today, we can study processes for more than a month (e.g. the hydration of calcium aluminates and silicates) the stability being such that fluctuations in the zero are less than 1/10th microwatt for almost unlimited time periods.

(ii) *The power produced, W, is constant:* This corresponds to the conditions employed in calibrating the galvanometer deviation, either by using the Joule effect, or by using a radioactive power source. The recorded graph, in the steady state, should be a straight line, parallel to the time-axis, at a height above the experimental zero which is perfectly reproducible and proportional to W.

(iii) *The power produced is variable:* This corresponds to the use of the apparatus as an oscillograph. The recorded curves must be reproducible for a given variation in power generated, and capable of correction as described in Chapter IV.

(b) Factors affecting the fidelity of the apparatus

(i) Causes of systematic errors

(a) *Irregular distribution of the thermocouples around the internal cell.* The e.m.f. produced in the detector pile must be proportional to the heat flow from the cell, and independent of the temperature distribution on the walls and within the internal cell.

(b) *Thermal leakage from the uppermost part of the cell, which is not covered by thermocouples.* Thermal leakage from this part of the cell must be made as small as possible, relative to the leakage from the walls. The cell stopper should be a thermal insulator, with a metallized surface. When using the Joule effect, the connecting wires must be in good thermal contact with the central metallic block (whilst remaining electrically insulated from it): however, it is preferable to use a radioactive source of constant power. The rods and tubes leading from the cell to outside the calorimeter must be of very low thermal conductivity.

(c) *Lack of identity of the twin calorimeters.* The calorimetric elements should be identical, and at the same temperature at any given time. The thermal perturbations can be divided equally between these elements (see p. 28)

(d) *Parasitic thermoelectric e.m.f.'s.* These can be totally suppressed by having all electrical circuits (including galvanometer and switches) enclosed within the calorimeter thermostat, close to the central block. This clearly cannot be done for high temperature operation of the calorimeter. One must, however, always thermostat the galvanometer and switches, and use carefully selected connecting wires of copper, enclosed in an insulated good-conducting metal tube.

(e) *Galvanometer faults.* The galvanometer must be insensitive to external magnetic fields, and the suspension wire completely free from residual torsion.

(ii) Accidental causes of error: There are several accidental causes of error, most of which are discovered by experience of working with the calorimeter. The most frequent among these are:

Evaporation loss of the liquid contained in the cell, due to defective stoppers;

Heat produced by mechanical effects (shock, vibration, adjustment of the connecting rods).

Bad electrical contacts. The currents in the detector pile are of the order 10^{-7} to 10^{-8} amps, so that it is essential that all electrical contacts should be properly made.
Humidity. A humid atmosphere can cause disastrous effects (galvanic currents, thermal effects due to condensation). The calorimeter must be operated at a higher

temperature than that of the room in which it is housed, and the air entering the room must be kept at a low level of humidity.

There are many other factors that can affect the reproducibility of the apparatus. However, a well-constructed microcalorimeter in the hands of an operator of average skill should, for the most part, prove completely reproducible in performance, and able to function at maximum sensitivity.

3. The Precision of Microcalorimetric Measurements

A detailed study of the precision obtainable with the microcalorimeter has been given in *Microcalorimétrie*. The overall precision depends on several factors, including the accuracy of calibration, the measurement of time and mass, the electrical measurements, the recording apparatus, and the methods of analysing the recorded curves.

We estimate that in measuring quantities of heat, a precision of 1 part in a thousand is attainable when compensation by the Peltier effect is made. For thermokinetic studies, the precision reached is usually of the order 1 per cent, but can surpass this in certain cases.

Conclusion

To sum up, the performance of the microcalorimetric apparatus, in regard to sensitivity, reproducibility, and precision, is sufficiently good to make possible its application to the study of a large number of processes over a wide and varied range in both physical-chemical and biological fields. The applications are described in the remaining parts of this book.

PART 2

PHYSICO-CHEMICAL APPLICATIONS

by E. CALVET

INTRODUCTION

THE apparatus described in Part I has already been employed to study a wide variety of different processes.

In many cases, a simple examination of the recorded curves has provided new factual data, and suggested new ideas. These curves generally give a detailed picture of the kinetics of the process under examination and often reveal features connected with it difficult to detect by other means.

For example, studies of adsorption processes in the microcalorimeter enabled us to follow the actual formation of a monomolecular film, since a marked discontinuity in the recorded curve occurs when the formation of the film is complete.

Studies on solvent-adsorption by soluble substances has revealed that a heat evolution occurs prior to the onset of dissolution: this seems to be a general phenomenon.

The hydration of calcium aluminates and silicates has been shown to proceed through a succession of chemical changes inducing successive crystallizations.

Microcalorimetric recordings made during spontaneous or induced precipitation from sodium aluminate solutions have enabled us to define the most suitable conditions for precipitation, and to suggest a mechanism of precipitation.

Direct thermochemical measurements of the heats of organic reactions have been made: the results so obtained are frequently of higher accuracy than can be obtained indirectly from bomb-combustion calorimetric studies.

Studies of catalysts at high temperatures have provided valuable information on catalytic activity and the mechanism of catalysis.

It is our belief that the microcalorimeter will shortly find its place as an instrument of fundamental importance in both research and industrial laboratories.

The following chapters describe the principal experimental studies made in our laboratory. They represent, however, only a beginning to the potential uses of the microcalorimeter in relation to physico-chemical and other fields of study.

CHAPTER 7

PHYSICAL MEASUREMENTS: SPECIFIC HEATS; THERMAL DIFFUSION COEFFICIENTS; HEATS OF SOLUTION, DILUTION, MIXING AND GELATINIZATION

1. The Measurement of Specific Heats

We shall describe the method in outline only: for a more detailed account, see *Microcalorimétrie*, p. 140.

The substance under investigation is contained in a cylindrical cell, top and bottom of which are closed by thermally insulated stoppers (Fig. 7.1). The cylinder and contents are uniformly heated by an electrical heater wound round the cylinder wall, and supplying a constant thermal power, W.

On switching on the heater, the laboratory cell reaches a steady thermal state after a short time interval. When thermal equilibrium is established, it can be shown that

(i) *the temperature is uniform within the cell,*

and (ii) *the equilibrium temperature, θ, is the same, for a given power, W, whatever the cell contents.* θ is the cell temperature relative to the external boundary.

When the thermal power, W, is cut off, the quantity of heat, Q_1, retained by the laboratory-cell is given by

$$Q_1 = (a + m_1 c)\,\theta \tag{54}$$

where m_1 = mass of substance placed in the cell, c = = specific heat of the substance, and a = apparent heat capacity of the container (cell plus heater).

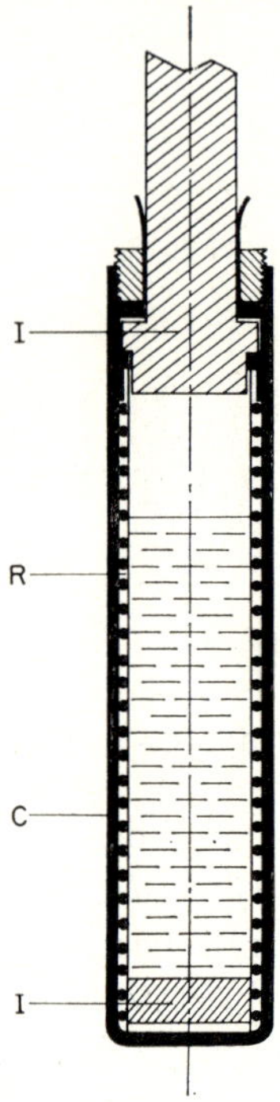

FIG. 7.1. Laboratory-cell for specific-heat measurements. C = cell; R = electrical heater; I = thermal insulation

Q_1 is measured from the area under the curve on its return to the experimental zero after cutting off the power, W. To determine c in (54), two additional measurements are made, keeping the thermal power, W, the same as before. Either, the original experiment is repeated twice, with different masses of substance in the cell, or, preferably, with masses m_2, m_3, of water in the cell; in this latter case, two further equations are obtained:

$$Q_2 = (a + m_2)\,\theta \tag{55}$$

$$Q_3 = (a + m_3)\,\theta \tag{56}$$

By solving the set of equations (54), (55), and (56) involving the three unknowns a, c, and θ, the value of c is then obtained.

The experiments may be carried out either by heating the cell and contents by electrical heating, as shown in Fig. 7.1, or alternatively by cooling the cell, making use of the thermoelectric pile for Peltier cooling. The method is capable of quite a high degree of accuracy.

2. The Measurement of Thermal Diffusion Coefficients, and of Thermal Conductivities

The accurate measurement of thermal conductivity is difficult, in contrast to the measurement of electrical conductivity. Methods which require temperature measurement can be particularly misleading, since the thermometer itself perturbs the temperature distribution around it, especially when good-conducting thermocouples are used: if direct contact between the thermocouples and the substance under investigation is prevented by insulating shielding, there is inevitably an appreciable temperature gradient across the insulation.

It is therefore preferable to measure thermal flux rather than temperature, or better, to compare the times required for a thermal shock to produce, under identical conditions, the same thermal perturbation in two different substances.

(a) The case of good conductors of heat

Suppose a laboratory-cell containing a liquid is heated (or cooled) by a constant power, W, generated at the cell wall by the Peltier thermopile. When thermal equilibrium is established, and the temperature difference between the internal and external boundaries is θ, then

$$W = p\theta = \frac{p}{g}\Delta$$

Let us now affix a cylindrical rod, *ab*, (made of the good-conducting material under investigation) along the cell axis so that the lower part is immersed in the cell liquid, and the top portion is in good thermal contact with the calorimeter block (Fig. 7.2).

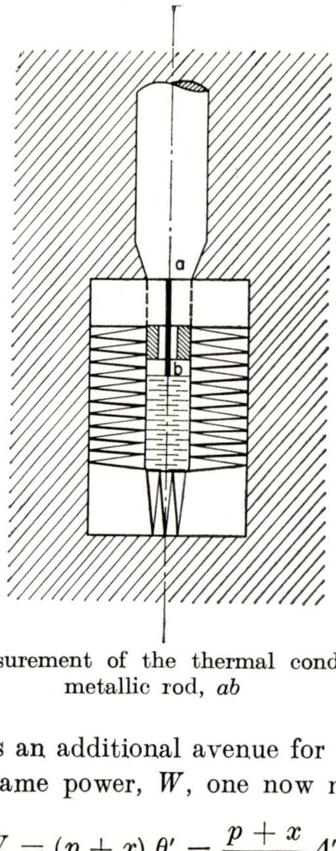

FIG. 7.2. Measurement of the thermal conductivity of a metallic rod, *ab*

The rod offers an additional avenue for thermal leakage, so that for the same power, W, one now may write

$$W = (p + x)\,\theta' = \frac{p + x}{g}\,\Delta' \tag{57}$$

where x is the thermal leakage along the rod, and θ' the temperature difference reached in the steady state.

Hence,

$$\frac{\Delta}{\Delta'} = \frac{p + x}{p}\,, \quad \text{and } x = p\left(\frac{\Delta}{\Delta'} - 1\right) \tag{58}$$

However, $x = K\frac{S}{l}$, where S = cross-sectional area of the rod, and l = length of the rod measured from the liquid surface to the point of contact with the calorimeter block. K is the thermal conductivity of the rod material.

(b) The case of bad conductors of heat

The method requires measurement of the heat of thermal disequilibrium in a cylinder of the material, heated along its axis.

A cylinder of the material under investigation is placed in the laboratory-cell of the calorimeter. This cylinder, of radius R, is pierced along its central axis by a hole of small radius r, through which passes a heating element producing power, W. When the steady state is established, there is thermal disequilibrium within the cylinder. It can be shown (*Microcalorimétrie*, p.145) that the coefficient of thermal disequilibrium, D, is given by:

$$D = \frac{R^2 - r^2}{4h} + \frac{r^2}{2h} \log \left(\frac{r}{R}\right) \qquad (59)$$

where h = coefficient of thermal diffusion $= \frac{K}{\gamma} = \frac{K}{c\varrho}$, K being the thermal conductivity, and γ the heat capacity of unit volume of the substance = specific heat × density = $= c\varrho$. Since r is small relative to R, the simplified equation $D = \frac{R^2}{4h}$ is a satisfactory approximation. The method of measuring $D = \frac{Q}{W}$ has already been described (p. 48).

3. The Measurement of Heats of Solution, Dilution, Mixing, and Gelatinization

The first measurements made with Tian's microcalorimeter (with compensation by the Joule and Peltier effects) were carried out by Mme. Bérenger-Calvet from 1925 to

1927, on the heats of solution and dilution of salts. Later, having improved the quality of performance of the apparatus, Mme. Bérenger-Calvet ventured the first microcalorimetric studies on slow processes (esterification and hydrolysis reactions), with commendable success.

Figure 7.3(1) shows the calorimetric device first used for heat of solution measurements. The cell, *C*, of thin glass and copper-coated, contained the solvent and two tubes, *T* and *J*. The glass tube, *T*, containing the solute, was tapered to a point at the bottom end, and attached to a rubber teat at the top. The second tube, *J*, oil-filled, contained an electric heater for compensation by the Joule effect. To start the experiment, the tube, *T*, was depressed to break the point, and the solvent was drawn up into the tube by manipulating the rubber teat. Tube, *T*, was then emptied and refilled by squeezing the teat several times until solution was complete.

Figure 7.3(2) shows the thermogram recorded during the measurement of the heat of solution of KNO_3 in water. In this particular experiment, the Joule compensation effect was initially overshadowed by the endothermic solution process, but later regained dominance, so that the "area correction" (see Chapter 4, p. 34) was very small.

The device shown in Fig. 7.3(1) was used by Mme. Bérenger-Calvet to make accurate measurements of the integral heat of solution of KNO_3, and of the heats of dilution from saturation down to $KNO_3.500\ H_2O$. Since these early experiments, we have made a large number of measurements of heats of solution and dilution, as well as measurements of heats of mixing, and heats of gelatinization. The type of mixing device used has varied according to the rate of solution and the volatility of the substances being studied. The apparatus in Fig. 7.3(1) is suitable for mixtures of non-volatile liquids: for studying the slow dissolution of nitrocellulose in acetone we used the device shown in Fig. 7.4(1);

that in Fig. 7.4(2) was used to study the solution of cellulose in a solution of cupric ethylenediamine salt.

These techniques are not satisfactory should significant variations in vapour pressure occur in the atmosphere above

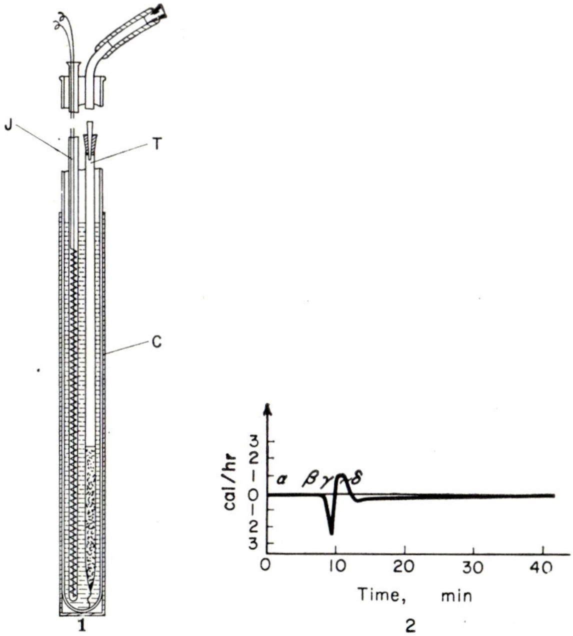

FIG. 7.3

1. Device used for measuring a heat of solution, with compensation by the Joule effect.
2. Thermogram recorded during a measurement of the integral heat of solution of KNO_3. An excess of KNO_3 was introduced into 0.805 g of water.

$\alpha\beta$ = the preliminary recording of the experimental zero;
β = the Joule heating current switched on;
γ = the tip of tube, T, broken;
δ = the Joule heating current switched off.

the liquid during the mixing or solution process. Pressure variations cause some vaporization or condensation of the liquid, the thermal effects of which cannot be disregarded. It is best to use a device such as that in Fig. 7.5, in which the vapour phase is eliminated from the experiment.

An interesting application of solution heat measurements has been in the analysis of the degree of crystallinity of

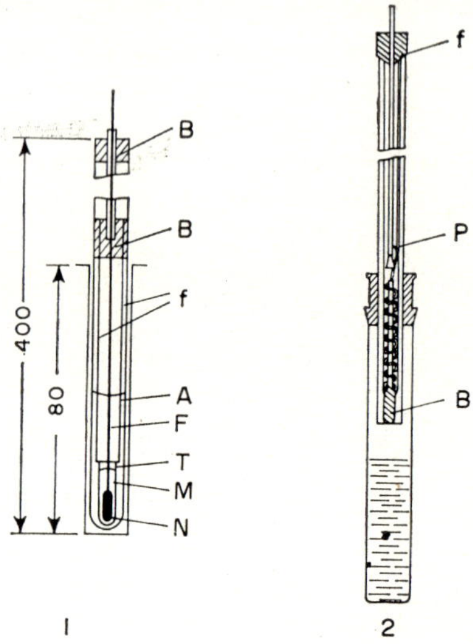

FIG. 7.4

1. Measurement of the heat of solution of nitrocellulose in acetone.
   B = cork stopper
   f = wire supporting the small tube, T, containing a bundle of nitrocellulose fibres immersed under mercury, M
   A = acetone solvent
   F = rigid steel wire.

By lifting the wire, F, the nitrocellulose fibres were brought into the acetone column.

2. Measurement of the heat of solution of cellulose in a solution of cupric ethylenediamine salt.

   A bundle of cellulose fibres is wound round a small glass tube sealed to the ground stopper, B. By depressing rod F, the stopper is released, and the bundle of fibres can be lowered into the solvent. The solution may be gently stirred by raising and lowering f.

alumina gel samples. It was shown by E. Calvet (1950) that the solution of alumina gel in soda solutions takes place in two stages:

(i) a rapid exothermic solution of amorphous alumina, and (ii) a slow endothermic solution of crystalline alumina.

Two stages are easily distinguished from the thermogram of the solution process, enabling the degree of crystallinity in a given gel sample to be readily determined.

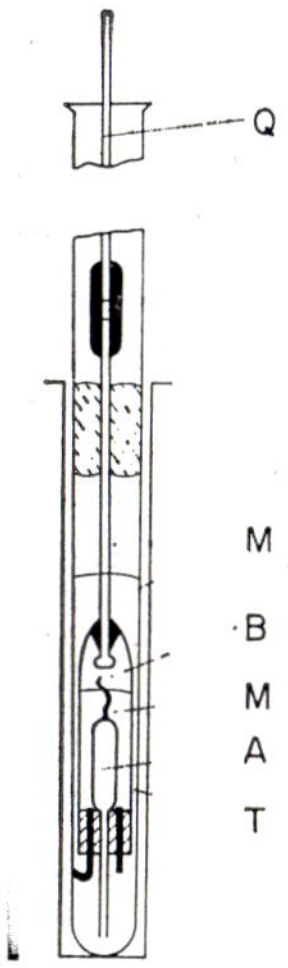

FIG. 7.5. Measurement of the heats of mixing of volatile liquids (e.g. alcohol-ether).

The liquids to be mixed are contained inside the glass tube, T, which can be moved from outside the calorimeter by the rod, Q. One of the liquids is sealed in the ampoule A, and the second held in the space, B, by the mercury seal, M. On depressing the rod, Q, the tapered point of ampoule A is broken; mercury displaces the liquid from the ampoule, so that the two liquids mix together in T without any vapour phase being formed.

E. Calvet and P.-H. Hermans (1951) have also estimated the degree of crystallinity in cellulose samples from measurements of their heats of solution in cupric ethylenediamine salt solutions.

CHAPTER 8

THERMOCHEMICAL MEASUREMENTS

1. General

For any given chemical transformation, one of the practical things to know is the value of the equilibrium constant. A direct determination of the equilibrium constant may not be feasible, and it becomes necessary to evaluate it from the heat and entropy changes associated with the transformation. The measurement of heats of chemical reaction is therefore one of the major tasks of experimental thermochemistry.

The heat of any given chemical reaction is obtainable from a knowledge of the *heats of formation* of the reactants and products of the reaction, since

$$Q = \sum Q_f(\text{products}) - \sum Q_f(\text{reactants}) \qquad (60)$$

where Q is the heat of reaction (positive if the reaction is exothermic), and Q_f symbolizes the heat of formation from elements ($Q_f = -\Delta H_f$) in defined states.

In practice, however, although extensive tables of standard heats of formation (at 25°C) have been compiled, and are continually being added to, the tables are by no means complete, and the data available are not always of high accuracy. Thus it may not be possible to calculate a desired heat of reaction because the requisite heats of formation are either lacking, or not known to a sufficiently high degree of precision.

The heats of formation of organic substances are normally obtained from heat of combustion measurements, many of which are nowadays made with high precision; but since combustion heats are often very large, the *absolute* error associated with a heat of formation determined by combustion calorimetry may still be appreciable. In the case of the heat of esterification of ethanol by acetic acid (Mme. Bérenger-Calvet, 1927) the heat of esterification is, in fact, smaller than the sum of the absolute errors associated with the available Q_f values of reactants and products. There are, therefore, distinct advantages to be gained from *direct* calorimetric measurements of reaction heats, particularly in cases where they are small.

Thermochemical studies of reactions are frequently carried out in some suitable solvent, so that the *observed* heat of reaction, Q, will differ from the *true* heat of reaction of the same substances in the pure state, and corrections must be made for the heats of solution of reactants and products in the solvent used. The true heat of reaction, X, is given by

$$X = Q + Q_1 - Q_2 \tag{61}$$

where X, Q, Q_1 and Q_2 refer to the following processes:

(i) $A + B \rightarrow C + D + X$
(ii) $A(\text{soln}) + B(\text{soln}) \rightarrow C(\text{soln}) + D(\text{soln}) + Q$
(iii) $A + B \rightarrow A(\text{soln}) + B(\text{soln}) + Q_1$
(iv) $C + D \rightarrow C(\text{soln}) + D(\text{soln}) + Q_2$

In the case where reaction (i) goes to completion, the corrections Q_1 and Q_2 are straightforward to apply; but in the case of incomplete reactions, the measurements of Q_1 and Q_2 must be made not in the pure solvent, S, but in a prepared mixture, S', composed of S and unchanged reactants. Suppose a fraction, α, of the reactants is used during the period of the thermochemical measurement of Q. Q_1 now represents the heat of solution of the fraction α of the reactants in a solvent, S', composed of the remaining fraction $(1 - \alpha)$ of the reactants in the pure solvent S. Correspondingly, Q_2 is the heat of solution in S' of the fraction α of the products. Since the solution processes are rapid relative to the reaction, the composition of the solvent medium remains unchanged during the time required for the heat of solution experiments.

2. Applications

1. The first direct microcalorimetric measurements of reaction heats were made by Mme. Bérenger–Calvet (1927), using Tian's microcalorimeter, who studied the *esterification* reaction

$$CH_3COOH + C_2H_5OH \rightleftharpoons CH_3COOC_2H_5 + H_2O \qquad (62)$$

in both the *forward and backward directions.*

The heat of esterification was found to be of the order + 1000 cal/mole: it may be remarked that the uncertainty in the published heat of combustion of ethyl acetate is itself larger than the heat of esterification.

For details of these studies, see *Microcalorimétrie*, pp.168-171.

2. E. Calvet and J. Dhers-Pession (1950) have made an analogous study of the *saponification of methyl nitrate,* from which they deduced that the heat of esterification of methanol by nitric acid is approximately 2000 cal/mole.

3. *The esterification of cellulose by nitric acid.* We have succeeded in establishing the thermokinetics of the esterification of cellulose by nitric acid–sulphuric acid mixtures, and also by nitric acid–acetic acid mixtures (*Microcalorimétrie,* p. 175–185). An interesting feature of these studies was the magnitude of the solution heat corrections: for instance, in a solvent mixture of composition by weight HNO_3, 14.5%, H_2SO_4, 70.7%, H_2O, 14.8%, the observed heat of reaction per C_6 unit of cellulose was 17,650 cal/unit,

of which only 2000 cal per —OH esterified came from the esterification reaction.

4. *Heats of saponification of amides.* The heats of saponification of a number of amides, and the thermokinetics of these reactions were studied by E. Calvet (Doctorate thesis, Paris, 1932). Figure 8.1 reproduces a thermogram obtained during the saponification of 1/200 mole acetamide by 1/200 mole caustic soda.

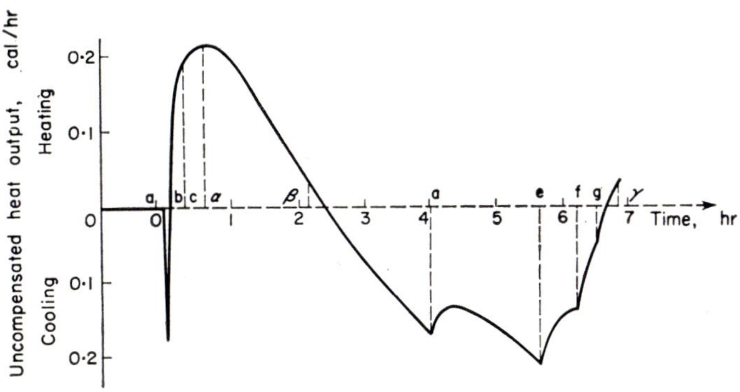

FIG. 8.1. The saponification of acetamide by caustic soda. Compensation by the Peltier effect.

a, *b* = cooling due to dilution of the soda solution
c = commencement of Peltier effect
d, *e*, *f*, *g* = cusps corresponding to changes made in the Peltier current
α, *β*, *γ* = times at which the amount of amide hydrolysed was determined.

5. From measurements of the heats of immersion of various samples of cellulose in caustic soda solutions, we have been able to find the best conditions for *mercerization*, and to determine the endothermic heat of mercerization.

The few examples cited above are ones in which we ourselves have been particularly interested. It is desirable that direct thermochemical measurements of reactions should be made on many more systems since the information obtained in this way is complementary to that obtained by combustion calorimetry, and indeed, by judicious usage of both calorimetric methods, it should be possible to build up more precise and extensive tables of chemical thermodynamic properties, of value to the organic chemist for the calculation of reaction conditions and equilibria.

CHAPTER 9

THE MICROCALORIMETRY OF ADSORPTION PROCESSES

1. General

The adsorption of a gas or vapour on a solid surface may be a purely physical process, or may involve chemical combination (chemisorption). Both types of adsorption are exothermic, but the heat of chemisorption is much larger than that of physical adsorption and serves to distinguish between the two types.

The thermokinetics of adsorption may be studied by tracing the thermograms produced when the vapour from a liquid, L, contained in the laboratory-cell is allowed to come in contact with the adsorbent, S, contained within the same cell. Under these conditions, the overall heat output is the heat of adsorption of the *liquid* transferred to the adsorbent.

We shall distinguish between the following:
(i) the adsorbent S is insoluble in the liquid L; (ii) the adsorbent is soluble in the liquid; and (iii) the adsorbent forms a gel with the liquid.

2. The Adsorbent is Insoluble in the Substance Adsorbed

Figure 9.1 shows the laboratory-cell as set up for an adsorption heat measurement. The cell contains the liquid, L, and a small cylindrical adsorption tube, F, in which is placed the adsorbent, S. The tap, R, in the neck of the tube can be manipulated by a connecting rod from outside the calorimeter: on turning tap, R, the vapour from L enters F and becomes adsorbed by S.

The tube, F, is removed from the cell and weighed at regular intervals during the adsorption experiment, and

FIG. 9.1. Laboratory-cell with fitments used for heat of adsorption measurements.

the increase in weight, m, plotted as a function of time of adsorption, $m = \Phi\ (t)$. The plot of the derivative, $\frac{dm}{dt} = \Phi'(t)$, taken in conjunction with the thermogram of the adsorption process, $\frac{dq}{dt} = f(t)$, enables the determination of the *differential heat of adsorption*, $\frac{dq}{dt}$ as a function of m.

This latter provides valuable information regarding the nature, physical or chemical, of the adsorption process, and is also useful in defining the activity of catalytic surfaces. It is desirable that it should be accurately measured, and to this end we introduced (E. Calvet, 1953) a microcalorimetric method of continuous weighing in order to determine the weight of substance adsorbed.

The method uses two laboratory-cells (Fig. 9.2), one containing the liquid and the other the adsorbent. The cells are connected through stopcock, R, which can be regulated from outside the thermostat.

When tap, R, is open, liquid distils from C_1 to C_2, causing C_1 to cool and C_2 to warm up. The cooling in C_1 is measured by the second thermopile (normally used for Peltier compensation) connected to a second galvanometer, and joined in opposition to the corresponding thermopile of

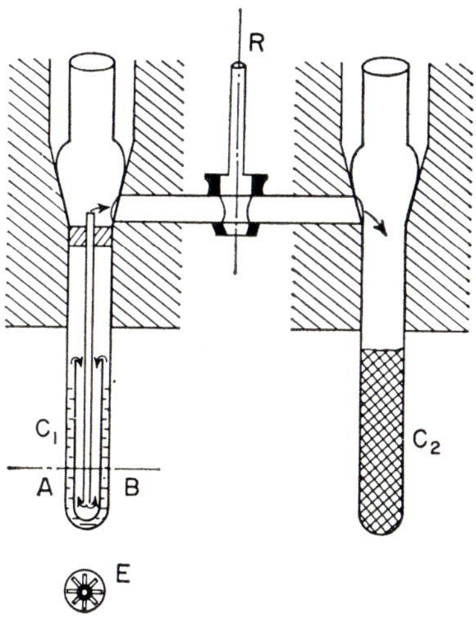

Fig. 9.2.

a reference element. *The recording given by the second galvanometer is a direct measure* of $\frac{dm}{dt} = \Phi'(t)$. To ensure that the distillation from C_1 is isothermal, we have used a silver heat exchanger, E, inside cell, C_1, and, having the form of a radiator with vanes.

(a) Measurement of the specific surface of powders

Thermokinetic studies of the adsorption by solids of substances in which they are insoluble have led to a somewhat unexpected application, namely, the measurement of the specific surface of the adsorbent, provided that this surface exceeds 5 m^2/g, and that the adsorption process occurs at a sufficiently slow rate. Under these conditions, the recorded thermogram shows a sharp discontinuity (change in slope) which corresponds to the completion of the formation of a monomolecular film on the surface of the adsorbent (E. Calvet, 1951 and 1953). Slow adsorption is necessary in order to avoid the formation of multimolecular layers on those parts of the adsorbent surface which happen to be more directly accessible to the vapour; time is required for diffusion over the whole available surface.

By measuring the weight of adsorbent, and the weight of adsorbed material at the onset of the break in the thermogram, the specific surface, S, of the adsorbent may be derived from the formula

$$S = p \frac{N}{M} \sigma \qquad (63)$$

where p = weight in grammes of substance adsorbed per g adsorbent,
M = molecular weight of adsorbed substance,
N = Avagadro number = 6.06×10^{23},
σ = area occupied by a molecule of adsorbed substance.

The sensitivity of this method naturally depends upon the accurate weighing of the adsorbed monolayer; as already indicated, this can be done in a very satisfactory way by the thermogravimetric method.

Note: The classical method of measuring specific surface area of very fine powders is the B.E.T. method (after Brunauer, Emmett and Teller), based on the study of adsorption isotherms: in this method, multimolecular films may be formed, and the mass of an *hypothetical* monomolecular film covering the entire adsorbent surface is obtained by calculation. The microcalorimetric method, on the other hand, depends on the formation of an *actual* monomolecular film whose energy of adsorption is sufficiently high.

Reference should also be made to a calorimetric method of measuring the specific surface of powders introduced by Harkins and Jura in 1944, in which a degassed sample of the adsorbent is treated with the vapour to the point of adsorption saturation, and then immersed in a sample of the liquid itself, contained in a sensitive calorimeter. The energy liberated in the calorimeter corresponds to the disappearance of the adsorbed film, and from this result the surface covered by the film can be deduced.

Figure 9.3 reproduces the recorded curve for the adsorption of water on a finely divided hydrargillite sample, $Al_2O_3 \cdot 3H_2O$: note the linearity of the plot preceding the sharp discontinuity at *A*. The linear portion indicates a constant heat liberation, obtained by careful adjustment of the stopcock, *R*, which controls the speed of adsorption.

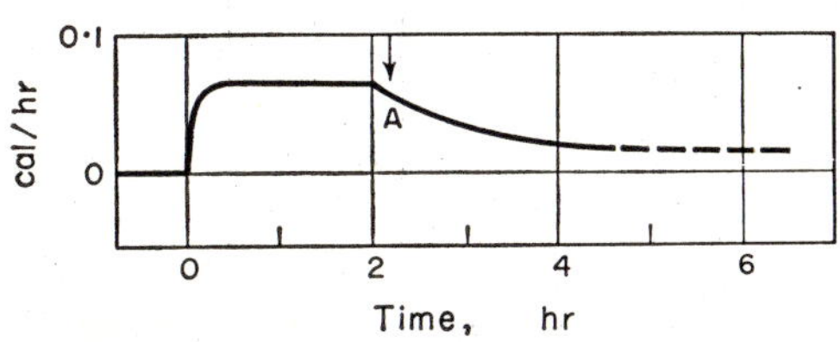

FIG. 9.3. Adsorption of water by powdered hydrargillite (0.110 g). At the point *A*, it had adsorbed 0.3% water.

We have made many measurements of specific surface area (see *Microcalorimétrie*, pp. 235–239). The results obtained agree excellently with those determined by the B.E.T. method, and by the method of Harkins and Jura.

(b) Determination of molecular surface area

The determination of the specific surface, *S*, from equation (1) requires that the molecular surface area, σ, of the adsorbed substance be known: inversely, if *S* of a given adsorbent is known, equation (63) can be used to determine σ. To measure an unknown σ, one firstly obtains the value, S, for a given adsorbent, by adsorption on the surface of a substance of known σ, e.g. water, for which $\sigma = 10.6 Å^2 = 10.6 \times 10^{-20}$ m^2.

Several measurements have been made in this way, including the study by E. Calvet and Benjamine Astruc

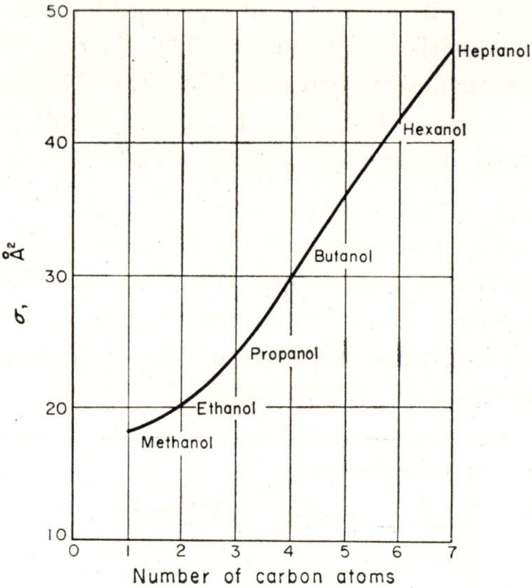

FIG. 9.4. Molecular surface areas (in Å^2) of the series of normal alcohols, C_1—C_7.

(1951) on the adsorption of the series of normal alcohols (C_1 to C_7) on a sample of hydrargillite powder of specific surface $S = 12.4$ m^2/g. The results are shown graphically in Figure 9.4.

The graph shows that from the third term to the sixth at least, the molecular surface area increases regularly by 6 Å^2 per CH_2-group added. Alcohol molecules do not therefore attach themselves normal to the surface as in the case of fatty acids on the surface of water, but lie flat along the adsorbent surface. The increase of 6 Å^2 in surface area corresponds to the projection in space occupied by a —CH_2 group in the plane including the C-atoms of the chain.

E. Calvet, H. Thibon and Josette Chapuis-Seite (1956) have measured the molecular surface area of acetone, using a sample of anastase of specific surface 11.3 m^2/g as adsorbent, and working at low temperature (3°C) in order to obtain a slow rate of adsorption. The result obtained (43 $Å^2$) is close to the approximate value given by calculation using "Leybold" atomic models.

3. Adsorption of Solvents by Soluble Substances

Our early experiments on the adsorption of solvents by solutes (E. Calvet 1938) were made to gain an insight into the *mechanism of dissolution*. To do this, we have tried to slow down the solution process by using an adsorption device similar to that shown in Fig. 9.1, in which the solvent vapour diffuses slowly on to the adsorbent surface. The recorded curves show the various thermal effects produced from the initial adsorption of the solvent molecules to the appearance of saturated solution.

(a) The adsorption of water on urea at 17.75°C.

Figure 9.5 shows the curves successively recorded during the adsorption of water by 1.80 g urea, initially dry. The different phases of adsorption were interrupted by intervals lasting for two days. The amounts of water adsorbed in the successive experiments I, II . . . VI were: 6.85 mg, 11.55 mg, 13.95 mg, 12.75 mg, 7.0 mg, and 6.5 mg.

These experiments were continued for 334 hr, when 148 mg of water had been adsorbed, without the appearance of a liquid phase. The following results were obtained:

1. The differential heat of adsorption, throughout the entire duration of the study, is constant and equal to 3600 cal/mole water adsorbed.

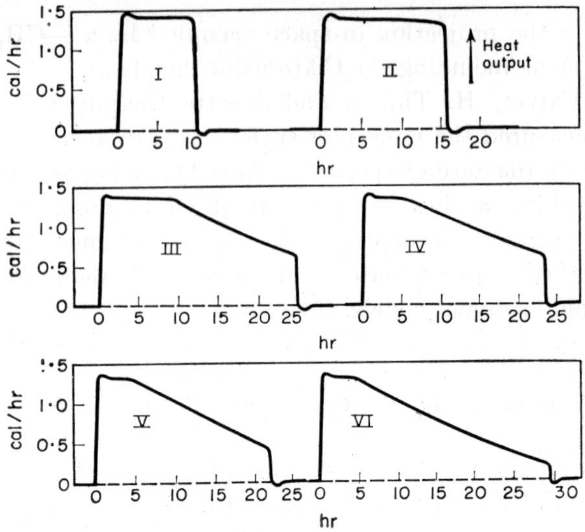

FIG. 9.5. Thermograms showing different stages in the adsorption of water by 1.80 g urea, with interruptions of 2 days between each stage.

2. The recorded curves each show an angular point which we interpret, as in the case of insoluble substances, as marking the completion of monomolecular film formation on the solid surface.
3. On closing the tap, *R* (preventing further adsorption), it seems there must occur a slow penetration of the water adsorbed on the surface into the crystal, and that this process is almost thermoneutral. Consequently, the crystal surface loses its monomolecular film, and on re-opening the tap, *R*, the surface adsorption initially observed recurs, with the same heat output as before.

(b) Other similar adsorption experiments

Similar experiments with different adsorbents and solutes have been made including acetamide and sucrose with water; benzamide with water, alcohol, and acetone; naphthalene with benzene, CCl_4, acetone, and hexane; I_2 with CS_2, etc. Analogous results to those of urea with water were obtained.

(c) The mechanism of dissolution

These experiments, followed all the way to solution of the crystal, show that dissolution is preceded by the exothermic adsorption of solvent molecules on the solute surface. The curves, $Q = f(n)$, giving the heat liberated by the adsorption of n moles of solvent on 1 mole of solute, always pass through a maximum, which may often represent a considerable heat evolution (Fig. 9.6). Following this, the endothermic breaking up of the crystal network takes place, solution occurs, and

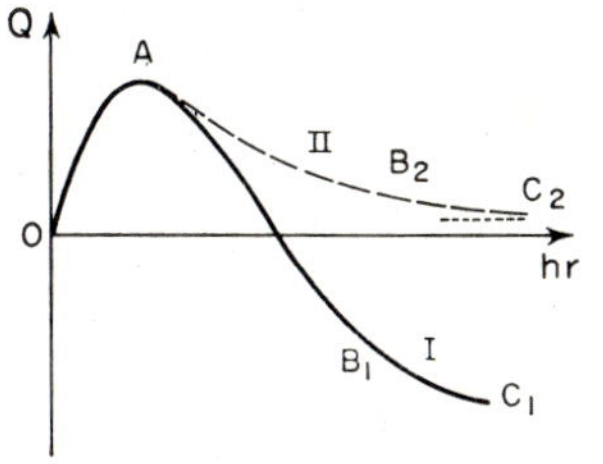

FIG. 9.6. Curves of $Q = f(n)$ obtained for solutes of average solubility,
OA = heating due to adsorption, preceeding the solution,
AB = cooling due to breakdown of the crystal lattice,
B_2C_2 or B_1C_1 = dilution of the saturated solution.

finally dilution of the initial solution: the total heat liberated is positive (curve AB_2C_2) or negative (curve AB_1C_1) depending on the degree of solvation of the dissolved substance.

It is more difficult to demonstrate the heat evolution prior to dissolution in the case of very soluble adsorbents, but we have succeeded in showing it in some cases, including those of phenol and diphenylmethane in acetone (see *Microcalorimétrie*, Chapter VIII, 2nd part).

CONCLUSION

Microcalorimetric studies have given a clearer picture of the mechanics of dissolution, showing that solvent adsorption precedes the appearance of a liquid phase, and that the quantity of adsorbed solvent may reach as much as 10% of the mass of adsorbent in favorable cases (e.g. urea and water).

4. Adsorption by Macromolecular Substances

(a) Adsorption of acetone by nitrocellulose

Figure 9.7 shows two thermograms obtained by the adsorption of acetone on a nitrocellulose sample. The two curves were recorded with an interval of two days between them. Note that the curves do not show a linear portion followed by a sharp break (as in Fig. 9.3), but that there is an almost exponential decay following the maximum. One is here faced with a prolonged adsorption process the mechanism of which is quite different from that of vapours on crystals.

(b) Physical and chemical adsorption

The differences in shape of the adsorption thermograms in Figs. 9.3 and 9.7 may be interpreted in terms of physical

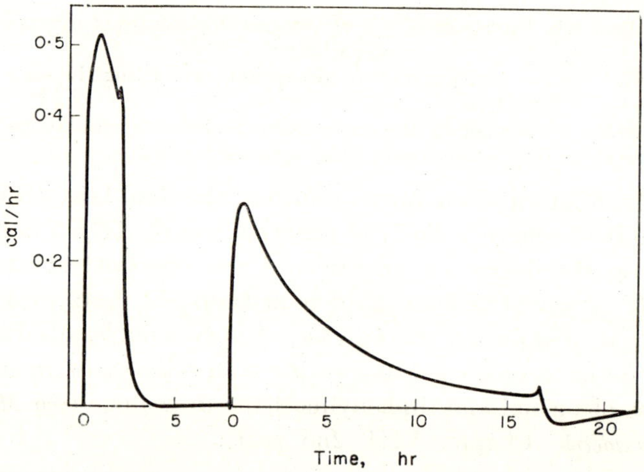

Fig. 9.7. Adsorption of acetone on a sample of nitrocellulose (directly recorded curves).

and chemical adsorption. When the rate of adsorption remains constant during the adsorption process despite the continuing reduction in free surface area due to adsorption, we conclude that the adsorbed molecules are *mobile with respect to the adsorbent surface,* being easily displaced by collision with on-coming molecules, thereby rendering the surface free to adsorb further. This type of adsorption, involving only a loose attachment of solvent molecules to the surface is referred to as *physical adsorption.*

Chemical adsorption, on the contrary, involves a firm attachment by chemical bonding of the solute molecules to the surface. The rate of adsorption is therefore proportional to the free surface area, and the kinetics follow an exponential law.

(c) Further evidence for the chemical adsorption of acetone by nitrocelluloses

Apart from the kinetics, which are characteristic of chemical adsorption, we might point out that the energy of adsorption of acetone by nitrocelluloses is high relative to the normal energies of physical adsorption processes.

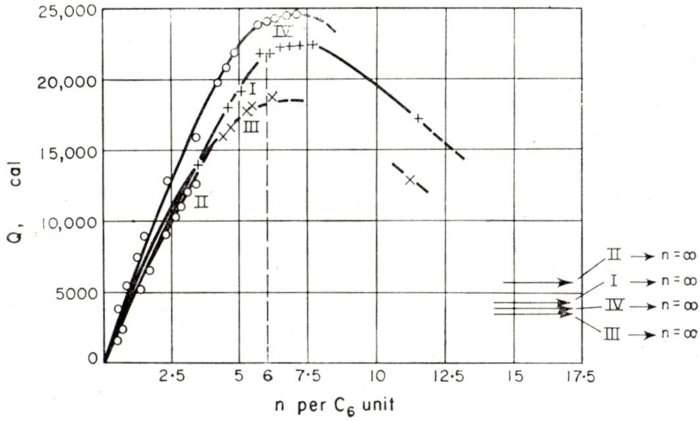

FIG. 9.8.

The graphs in Fig. 9.8 (E. Calvet, 1941–1942) show that the heat evolved during the slow adsorption of n moles acetone per C_6-unit of di- or tri-nitrocelluloses as fibre or film is approximately 6000 cal/mole for $\frac{dQ}{dn}$ at the commencement of adsorption. This figure is the same as that due to the formation of a hydrogen-bond, and suggests that the acetone molecules may attach themselves to hydrogen atoms in nitrocelluloses.

It was reported in 1942 that the initial heat of adsorption of acetone by a nitrocellulose is independent of the number of $—NO_2$ groups per C_6-unit, which is immediately understood if the adsorption occurs solely by hydrogen-bonding.

Furthermore, it should be noted that although the heat evolved by adsorption of acetone diminishes as n increases, due to the swelling of the fibres, heat continues to be evolved until n reaches a value of 7, at the flat maximum of the curve $Q = f(n)$: following this, acetone vapour no longer adsorbs on the cellulose surface. To trace the curve $Q = f(n)$ beyond $n = 7$, one can adsorb liquid acetone, which causes the gel to dissolve with cooling.

The coincidence between the number of acetone molecules adsorbed at the maximum of $Q = f(n)$, and the number of H-atoms per C_6-unit of cellulose is striking, and is strong evidence for hydrogen-bonding as the mechanism of adsorption in this case.

Further adsorption studies on macromolecular substances

In addition to the adsorption of acetone on nitrocelluloses, we have also studied the adsorption of acetate esters, of alcohol-ether mixtures, and of methyl nitrate on nitrocellulose samples; of acetone on bakelites; of various solvents on polyvinyl chloride, etc. In all cases, the curves $Q = f(n)$ pass through a maximum if the adsorbent is soluble.

These examples should be sufficient to show the value of calorimetric and thermokinetic measurements in the study of the mechanism of gelatinization of macromolecular substances, and of the composition and structure of solute-solvent complexes.

CHAPTER 10

SOME MICROCALORIMETRIC STUDIES ON THE CHEMISTRY OF ALUMINA

(E. CALVET and H. THIBON)

1. STUDIES ON THE BAYER PRECIPITATION

The precipitation of alumina trihydrate from solutions of sodium aluminate is one of the basic steps in the preparation of alumina by the Bayer process.

E. Calvet, P. Boivinet, H. Thibon and A. Maillard (1951) have carried out the Bayer precipitation in the laboratory-cell of the microcalorimeter under various conditions, in the absence of crystal nuclei, in the presence of aluminium hydroxide nuclei of diameter less than one micron, and in the presence of alumina gels.

(a) Spontaneous precipitation at 17°C in the absence of nuclei.

The following facts have been established:

(i) The aging of the solutions prior to precipitation taking place occurs without any accompanying thermal effects: the heat evolution is confined to the precipitation process.

(ii) The amount of heat liberated is approximately proportional to the weight of alumina trihydrate precipitated, i.e. *ca.* 110 cal/g Al_2O_3, and is independent of the composition of the solution.

(iii) X-ray analysis of the precipitate has shown it to contain two forms of $Al_2O_3 \cdot 3\,H_2O$, namely hydrargillite and bayerite. Since the heat of precipitation remains constant despite variations in the relative proportions of the two constituents, it follows that the transformation bayerite → hydrargillite is virtually thermoneutral.

(b) Precipitation at 35°C, induced by microcrystalline nuclei of aluminium hydroxide

The device shown in Fig. 10.1 was used to introduce crystal nuclei into 10 cc of a solution of aluminate, contained in the laboratory-cell. The three small nickel discs, each smeared

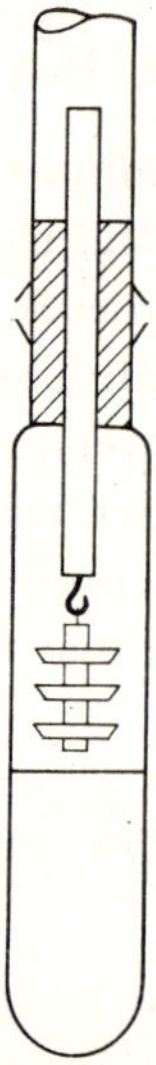

FIG. 10.1.

with a thin microcrystalline layer, were supported above the aluminate solution prior to the experiment: by depressing the shaft holding the discs, the crystal nuclei were totally immersed in the solution at the desired instant. The observations showed there are two main thermal effects, the first occurring immediately on immersion, the second later effect coinciding with the actual precipitation process.

At the instant of immersion, in the case where the nuclei were $Al_2O_3 \cdot 3\,H_2O$ crystals, there occurs a sharp thermal effect characteristic of an adsorption process, and due to the solvation of the surface of the crystal nuclei introduced. The sign of this thermal effect depends on the value of the molar ratio Na_2O/Al_2O_3 of the solution, and is zero for a particular value, R_0, of this ratio.

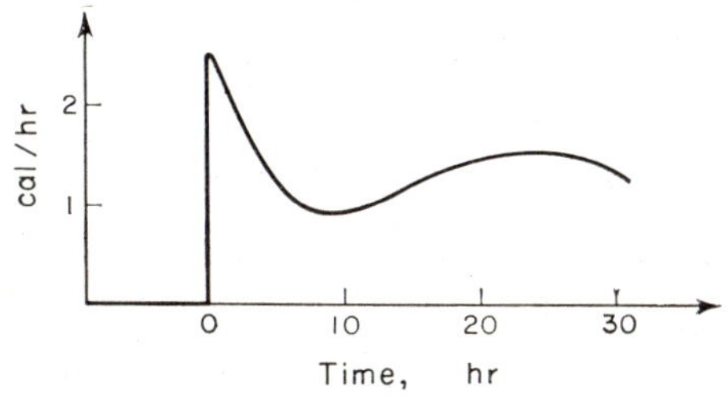

FIG. 10.2. Decomposition of a solution of industrial sodium aluminate after immersion of very finely powdered hydrargillite.

Following the initial effect, and some considerable time later, precipitation of trihydrate takes place, and is accompanied by heat evolution of the order 110 cal/g Al_2O_3 (see Fig. 10.2).

(c) Precipitation induced by alumina gels

The alumina gels, usually prepared by adding sulphuric or hydrochloric acid to a solution of sodium aluminate, were shown by X-ray analysis to consist of amorphous aluminium hydroxide often admixed with semi-crystalline boehmite (pseudoboehmite).

These gels invariably caused a sharp evolution of heat on immersion into sodium aluminate solution, coincident with gel solution: subsequently, the slow exothermic precipitation process took place, the two thermal effects usually being well separated from each other. The precipitate was hydrargillite mixed with a little bayerite, but no boehmite, and the heat of precipitation was again approximately 110 cal/g Al_2O_3. Finally, it has been established that the alumina gels are far more efficient than the microcrystalline trihydrates for inducing the Bayer precipitation.

2. Studies on Alumina Gels

(a) Solution in 4 N caustic soda

E. Calvet, H. Thibon *et al.* (1953), using the immersion device of Fig. 10.1 for introducing gels into 4 N caustic soda, have shown that the thermograms given by amorphous gels show only an initial exothermic response, whereas in the case of pseudoboehmite gels the initial exothermic response (due to solution of the amorphous part of the pseudoboehmite) was

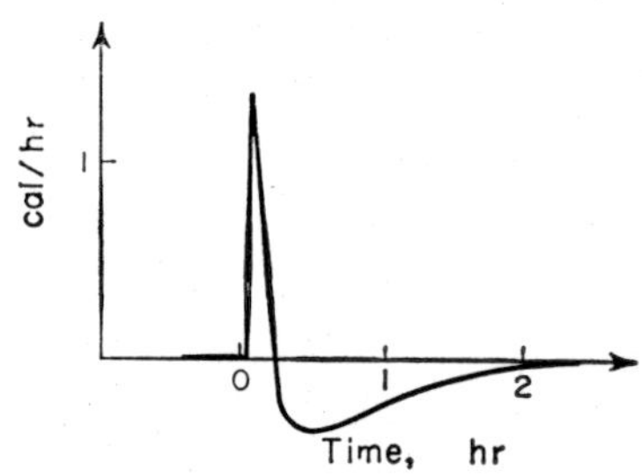

FIG. 10.3. Solution of a gel admixed with crystalline alumina in 4 N caustic soda.

followed by an endothermic effect, due to solution of the semi-crystalline pseudoboehmite (Fig. 10.3).

We can thus distinguish between an amorphous gel and a pseudocrystalline gel without the aid of X-ray analysis. The heat of solution of gels in 4 N soda has been shown to depend on the degree of crystallinity of the gel, on their occluded anions (gels prepared by sulphuric acid, or hydrochloric acid neutralization of aluminate), and on the temperature at which the gels were dried.

(b) Solution in 2 N sodium carbonate

E. Calvet and H. Thibon (1953) have carried out a similar series of studies on the solution of gels in 2 N Na_2CO_3 solution: the results were quite different from those found with 4 N NaOH as solvent. The thermograms obtained on solution of amorphous gels showed a strong initial fall, due to an endothermic process (Fig. 10.4), whilst the thermograms given by pseudoboehmite gels showed only a slight kink, also endothermic.

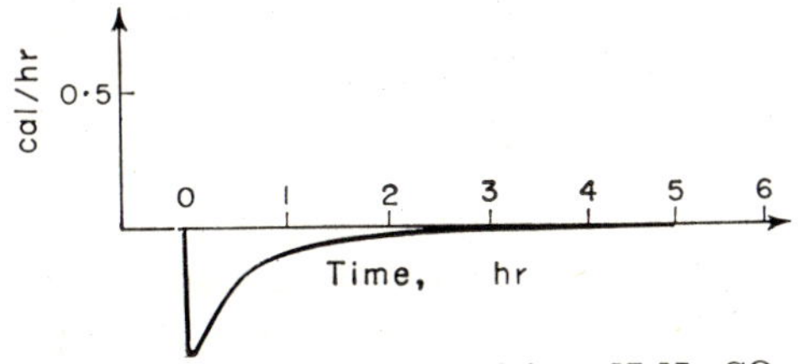

FIG. 10.4. Amorphous gel immersed in 2 N Na_2CO_3 solution.

Detailed chemical analysis of the solution process has shown it to involve two successive reactions:

(i) $Al_2O_3 + Na_2CO_3 \rightarrow 2NaAlO_2 + CO_2$ (endothermic)

and (ii) $2NaAlO_2 + 2CO_2 \rightarrow Na_2O.Al_2O_3.2CO_2$ (exothermic)

The second stage is too slow and incomplete to be detected in the microcalorimeter: thus the thermograms recorded only the first endothermic stage.

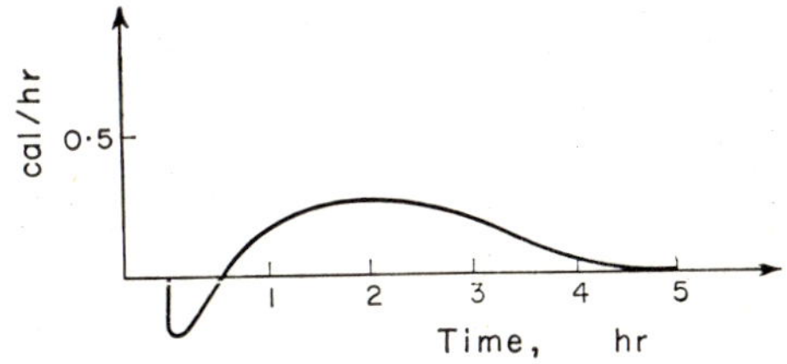

FIG. 10.5. Amorphous gel immersed in a 2 N solution of a mixture containing 73% Na_2CO_3 and 27% $NaHCO_3$.

The second stage can be accelerated and intensified by using as solvent a solution of Na_2CO_3 containing a certain proportion of $NaHCO_3$: the thermogram in Fig. 10.5, obtained on solution in a mixture of carbonate and bicarbonate, shows the broad exothermic bulge, following the initial endothermic fall, corresponding with the precipitation of sodium aluminium double carbonate.

3. Microcalorimetry of "Activated" Alumina

Microcalorimetric methods have proved very useful in the study of the so-called "activated" or "active" aluminas, which are formed by controlled thermal decomposition of the trihydrate.

(a) The determination of the threshold temperatures for decomposition of different varieties of hydrargillite

E. Calvet and H. Thibon *et al.* (1957), in studies on the thresholds of thermal decomposition of different varieties of hydrargillite (differing in texture, coarseness or fineness of the granules) have used the microcalorimeter to identify activated alumina in the products of decomposition, and thus verify that the threshold temperatures were reached.

The apparatus for measuring specific surface area of powders, (Fig. 9.1, Chapter 9), was used in these experiments which have clearly demonstrated that the thermal decomposition thresholds are dependent on the textures of the hydrargillite samples (the decompositions were carried out at atmospheric pressure, in a stream of dry air). For instance, normal Bayer hydrargillites (spherical granules of mean diameter about 40 microns) do not decompose below 145–150°C, whereas a very fine hydrargillite powder, with a specific surface 300 times greater, commenced to decompose at *ca.* 130°C. This difference is probably due to the far greater difficulty of diffusion of the water vapour formed to the surface of the powder in the case of the coarser Bayer hydrargillites.

(b) The variation of specific surface of activated aluminas with temperature of activation

E. Calvet, H. Thibon and J. Chapuis–Seite (1956) used the calorimetric method to measure the specific surface areas of five alumina samples prepared by heating ordinary Bayer hydrargillite (granules of diameter *ca.* 40 microns, for 48 hr, at atmospheric pressure, at 200, 250, 300, 350, and 400°C respectively. Using acetone vapour as solute they found a continuous increase in specific surface as the temperature of activation increased from 200—400°C. The increase in specific surface coincides with diminished retention of water by the activated alumina, and is due to the growth of a porous network within the crystals which confers the adsorbent properties. The specific surface depends upon the accessibility of the pores to the adsorbed molecules, and therefore on the mean diameter of the pores and the molecular surface area of the adsorbed substance.

(c) The rehydration of activated alumina

D. Papée, R. Tertian and J. Charrier (1953, 1955) have investigated the conditions under which activated alumina will, by contact with water at ordinary temperature, rehydrate to form bayerite, $Al_2O_3.3H_2O$. The important conditions for rehydration are that the alumina should be activated by heating at 180–210°C under low pressure (10^{-2} mm Hg), and that very finely powdered (*ca.* 1 micron diameter particles) trihydrate be used initially.

E. Calvet and H. Thibon (1955) have studied the thermokinetics of hydration of active aluminas at 35°C. They used aluminas prepared from very finely powdered hydrargillite (diameter *ca.* 1 micron) by heating at temperatures in the range 180–500°C under atmospheric pressure in some cases, and low pressure (10^{-2} mm Hg) in others.

The thermograms of hydration of these alumina samples differed according to the conditions of activation of the samples. Vacuum-activated aluminas gave thermograms showing a sharp initial exothermic rise, due to adsorption (wetting of the surface), followed by a bell-shaped exothermic bulge corresponding to chemical hydration and the formation of a crystal phase identified by X-ray analysis as bayerite: but the thermograms given by aluminas activated at atmospheric pressure showed only the initial surface-wetting effect, and no bayerite was formed from these samples.

Comparison of the thermograms given by vacuum-activated aluminas has also shown that it is disadvantageous to activate at too high temperatures; aluminas activated in the temperature range 205–520°C were far less readily hydrated than those activated from 180 to 205°C. Thus, the microcalorimetric studies have fully confirmed the findings of Papée, Tertian and Charrier by other methods.

CHAPTER 11

MICROCALORIMETRY APPLIED TO THE CHEMISTRY OF CEMENTS

1. The Hydration of Calcium Aluminates
(E. Calvet and P. Longuet, 1954)

The hydration studies were made in similar fashion to those on the hydration of activated aluminas, using samples of aluminates prepared at the Centre d'Etudes et de Recherches de l'Industrie des Liants Hydrauliques. Studies were made on four aluminates, A_2C, AC, A_7C_{12}, and AC_3, and on mixtures of these (the symbolism used is commonly employed in cement chemistry; $A = Al_2O_3$, $C = CaO$, and $S = SiO_2$).

The recorded curves (Fig. 11.1) were generally of the same characteristic shape having two maxima corresponding to two distinct exothermic processes.

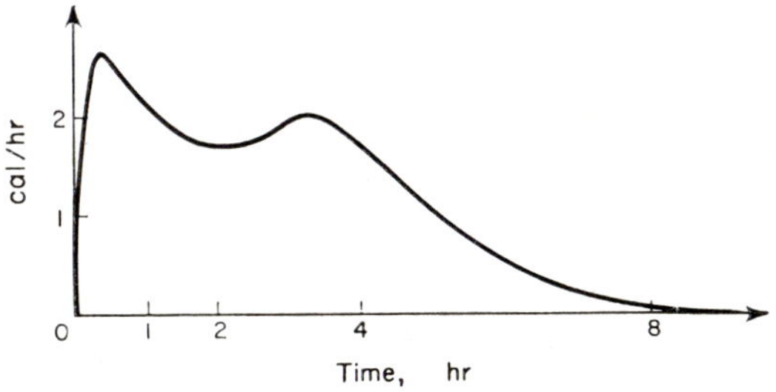

Fig. 11.1. Hydration of calcium aluminate, A_7C_{12}.

The shape of the curves in detail, depends on the nature of the aluminate used and the temperature, the two thermal processes being more clearly separated the higher the ratio Al_2O_3/CaO, and the lower the temperature.

In the following table are listed the heats liberated during the hydration of pure aluminates: in some cases, where the two maxima of the curves were well separated, it was possible to measure the heat liberated in each part of the hydration process.

TABLE 1

Aluminate	Temp.	Duration of experiment	Heat liberated (cal/g aluminate)		
			1st part	2nd part	Total
A_2C	25	9 days	2.2	70	72.2
AC	35	24 hrs	4.3	125	129
AC	10	4 days	7	86	93
A_7C_{12}	35	26 hrs			160
A_7C_{12}	10	29 hrs	7.2	86.5	94
AC_3	25	18 hrs			300
AC_3	10	22 hrs			300

From these results, it would seem that the richer the aluminate in CaO, the more important relatively the heat liberation in the first part. However, the solid separated by filtration after the first part showed no trace of hydrate: hence, we interpret the first thermal effect as due to solution of the anhydrous material, and the second to the formation of crystalline hydrates.

In a more recent study, we have observed that "metastable" solutions of calcium aluminate decompose exothermically: the thermogram has the same shape as the second part of the hydration curves.

2. THE HYDRATION OF CALCIUM SILICATES

The thermograms obtained by hydration of di- and tri-calcium silicates show the same characteristic form of those given by calcium aluminates: there is an initial sharp exothermic effect followed by a second exothermic stage giving a bell-shaped curve (Fig. 11.2).

The heats evolved were considerably less than with aluminates, enabling us in the case of tricalcium silicate to study the effect of varying the ratio SC_3/H_2O on the exothermicity of the process.

The two exothermic parts were more clearly separated from one another than in the case of aluminates: the first

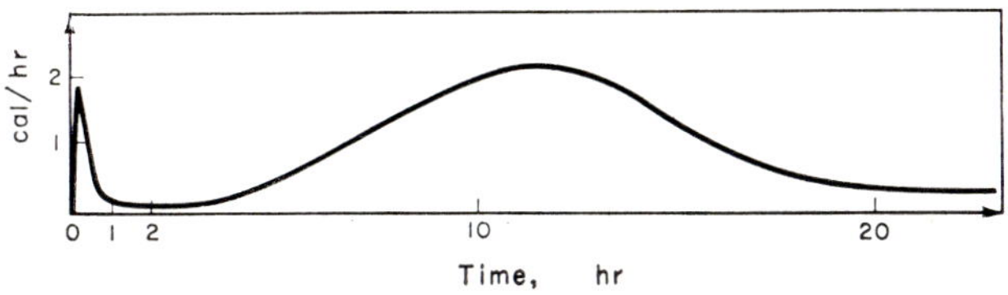

FIG. 11.2. Hydration of tricalcium silicate, SC_3.

part corresponds to the formation of saturated solution, the second to the formation of solid hydrates. The heat evolved in the first part was proportional to the quantity of lime present in the final solution, and in the second part to the quantity of anhydrous product used.

TABLE 2

SC_3/H_2O molar ratio	CaO dissolved (g/l)	Heat evolved	
		1st part (cal/g CaO dissolved)	2nd part (cal/g silicate)
1/10000	0.20	85	38
1/1000	0.40	83	36
1/100	1.3	77	33
1/25	3.1	85	36

We have also studied the hydration of industrial cements: the thermograms again showed the two stages characteristic of the hydration of silicates and aluminates, but the rate of hydration was very much less.

CHAPTER 12

MICROCALORIMETRY AT ELEVATED AND AT HIGH TEMPERATURES

1. Description of the Microcalorimeter for Operation at Elevated Temperatures

It has been necessary to make a number of modifications to the ordinary Calvet microcalorimeter in order to adapt it for operation at elevated and high temperatures (100°–1000°C). The modified calorimeter is shown in Fig. 12.1.

The two cells are made from platinum. The substance under investigation is placed in a Pyrex test-tube, which fits closely inside the laboratory cell. An identical Pyrex tube is placed in the reference cell: this may be packed with glass beads of volume equal to the contents of the laboratory cell.

Each cell is surrounded by a thermoelectric pile built up from twelve coils containing twelve Pt-to-Pt/Rh thermocouples (144 couples in all). The two piles are connected in the differential arrangement, and the differential e.m.f. recorded by the deviations of a spot-galvanometer.

The cells are contained in symmetrically disposed cavities inside a steel block: to ensure equipartition of the thermostat fluctuations between the two cells, the system of cones above and below the block is employed, as described previously (p. 28). The assembly is placed in a steel jacket, surrounded by a cylindrical muffle of refractory cement

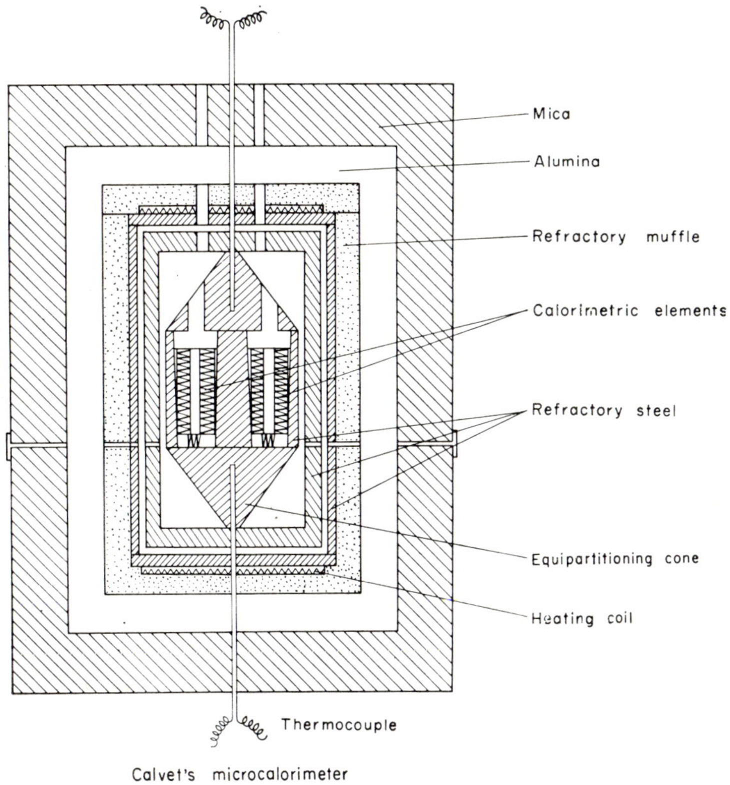

FIG. 12.1. Microcalorimeter for use at elevated temperatures.

and several concentric jackets insulated from one another by mica or alumina. The heating of the block is by two heating coils in the base of the largest steel jacket, and controlled by an electronic thermo-regulator.

2. Method of Using the Apparatus

The calorimeter is particularly suited to the study of gas-solid reactions at elevated temperatures. The two cases of interest are:

(i) the solid decomposes on heating, producing a gaseous product—e.g. dehydration of crystal hydrates;

(ii) the solid on heating becomes active as a catalyst, promoting reaction in a mixture of gases in the cell.

In both cases, the laboratory cell is charged with the solid, the calorimeter assembled, and the temperature of the block raised to a suitable level and held stable for a sufficient period to establish the constancy of the experimental zero. The reaction is then "started"—either by evacuating the laboratory cell, or by admitting the gas mixture into the cell. We shall describe examples of both types of experiment.

3. Dehydration Studies

Max Michel (1955–6) studied the thermal dehydration of gypsum, and later of three forms of hydrated alumina (namely, the trihydrates hydrargillite and bayerite, and the monohydrate, boehmite.) From 0.5 to 1 g of powdered hydrate was placed in the reaction tube of the laboratory cell (generally a quartz tube was used), and the microcalorimeter raised to a convenient starting temperature (90° for gypsum, 230° for hydrargillite, 210° for bayerite, 500°C for boehmite). The reaction tube was then evacuated. The vacuum line included a set of P_2O_5 adsorption tubes, interposed between the valve attaching to the exit tube from the reaction cell and the pump. The progress of the dehydration reaction in the calorimeter was followed by recording the gradual increase in weight of the adsorption tubes as a function of time.

Comparison of the thermogram recorded during the dehydration process with the curve obtained by plotting the weight increase of the adsorption tubes with time enabled a quantitative evaluation of the heat of dehydration of each of the hydrates studied. The results are summarized in Table 1 below: the measured heats (from the areas under the thermograms) refer to the temperature of the experiments and the low pressures employed. They are corrected in Table 1 to standard conditions (20°C, 760 mm pressure).

TABLE 1

Compound	Gypsum $CaSO_4 \cdot 2H_2O$	Hydrargillite (very fine) $Al_2O_3 \cdot 3\,H_2O$	Bayerite (fine) $Al_2O_3 \cdot 3H_2O$	Boehmite (very fine) $Al_2O_3 \cdot H_2O$
Temp. of dehydration (°C)	90	230	210	500
Pressure (mm Hg)	1	10^{-2}	10^{-2}	10^{-2}
Duration of experiment (hrs)	15	10	10	6
Final product	$CaSO_4$	Al_2O_3, ϱ mixed with a little Al_2O_3, χ or η but no boehmite.		Al_2O_3, γ
ΔH of dehydration (kcal/mole) at 20°C, 760 mm	25.7	48.0	46.5	17.0

These results have been verified in the case of gypsum and bayerite by direct measurements of the heats of hydration of anhydrous $CaSO_4$ and alumina in the normal Calvet calorimeter, operated at 35°C.

4. STUDIES IN CATALYSIS

The high temperature calorimeter can be used to study several features of heterogeneous catalysed gas reactions, and to assist in clarifying the mode of action of the catalyst. In these experiments, the laboratory cell is charged with a known weight of catalyst and filled with a mixture of the reactant gases. The temperature of the calorimeter assembly is progressively raised and the thermogram continuously recorded: the onset of reaction is revealed by a marked response in the thermogram, enabling the temperature range of catalyst activity to be determined.

Investigations may also be made at *fixed temperatures* within the range of catalyst activity. For instance, one can measure the thermal effect produced by each gas separately on adding it to the catalyst: in this way it is possible to decide whether one of the gases in the reaction mixture is preferentially adsorbed on the catalyst surface relative to the others. One can also vary the *composition* of the gas mixture introduced into the cell, to find the most satisfactory mixture: correspondingly, the effect of changes in the physical state or composition of the catalyst can readily be investigated.

The oxidation of ethylene

As an example, we may refer to a series of studies made on the controlled oxidation of ethylene, to form ethylene oxide.

$$C_2H_4 + {}^1/_2\,O_2 \rightarrow \underset{\diagdown\;O\;\diagup}{CH_2 - CH_2} \qquad (64)$$

This reaction is carried out industrially by reacting a mixture of 10% ethylene and 90% air on silver oxide catalyst at 270°C. The microcalorimetric studies were therefore restricted to this temperature (270°C), and to the use of the industrial catalyst. In each experiment, the cell contained 1 g catalyst, and thermograms were recorded on treatment with air, with ethylene, with each gas in turn, and with the mixed gases. The recorded thermograms are reproduced in Fig. 12.2.

(i) *Action of air.* The addition of air produced a very sharp exothermic peak, followed by a rapid return to the level of the experimental zero. After evacuating the cell (represented by *v* in the Figure), a further air treatment produced a second peak, but of less intensity. The peak heights fell off with each such air treatment, from 17 cm, initially to 5 cm after the third admission of air. On evacuation, a small endothermic peak (1.7 cm in height) was noticed. After several additions of air, ethylene was introduced into the cell: there resulted a sharp exothermic peak rising to 17 cm, similar to that observed on the first addition of air. The peak was followed by a rapid fall to the zero-line. Addition of air subsequent to the addition of ethylene gave once more a sharp

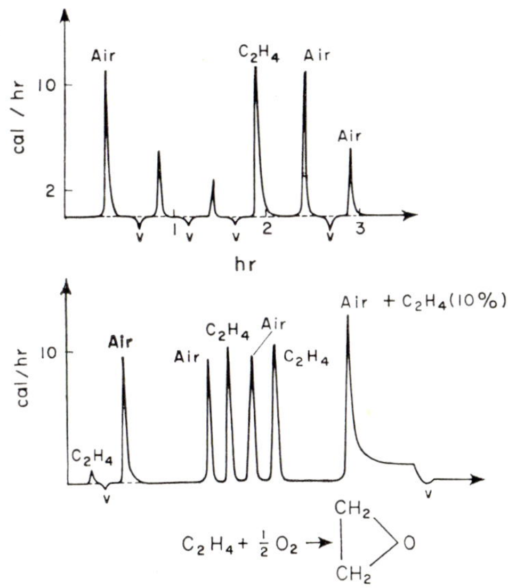

FIG. 12.2. Thermograms produced on adding air, ethylene and mixtures of the two gases to silver oxide catalyst at 270°C.

peak, rising to 17 cm: thus the effect of adding ethylene is to revive the activity of the catalyst towards air. The simple explanation is that the adsorption of air by the catalyst is in part *physical* adsorption (this part being removed by evacuation), and in part *chemisorption* (this part being retained on evacuation, but lost when treated with ethylene, which reacts with the chemisorbed oxygen).

(ii) *Action of ethylene.* The treatment of a fresh catalyst sample with ethylene produced only a small exothermic peak (1.7 cm), and after evacuation, further ethylene addition gave only 0.7 cm peak response. Admission of air (after evacuation) resulted in a sharp peak 17 cm high. Alternate treatments with ethylene and air produced a series of sharp peaks. Thus it seems clear that ethylene is not adsorbed by the catalyst, *unless this already carries a layer of adsorbed oxygen.* In this event, the ethylene is retained by chemical reaction with the adsorbed oxygen on the surface.

(iii) *Action of an air/ethylene mixture.* A mixture of air and ethylene (10% C_2H_4) was admitted to the cell at 700 mm pressure. The thermogram showed a sharp initial rise (20 cm), falling quickly to a level which remained steady 1.5 cm above the zero-line. The steady reaction continued for a prolonged period, during which the pressure in the reaction vessel diminished, as would be required by equation 64. On evacuating the cell, the thermogram returned instantly to the experimental zero-line.

Finally, measurements were made of the heat evolved on reacting the air-ethylene mixture under identical conditions with different catalyst samples. It was shown that on a fresh catalyst sample, heat was evolved at 3.35 cal/g/hr; on a well-used catalyst sample, the heat output fell to 1.39 cal/g/hr. Thus it was possible, by microcalorimetric studies, not only to gain insight into the mode of action of the catalyst, but also to show that the efficiency of the catalyst deteriorates markedly with continued usage.

BIBLIOGRAPHY OF PARTS 1 AND 2

1923 A. TIAN; Microcalorimeter with compensation by Peltier and Joule effects. *Bull. Soc. Chim. Fr.*, **33** (4), 427.

1924 A. TIAN; *C. R. Acad. Sci., Paris*, **178**, 705.

1925 Marcelle BERENGER and A. TIAN; Heats of solution, and heats of incomplete reactions. *C. R. Acad. Sci., Paris*, **180**, 1907.

1927 Mme BERENGER-CALVET; Compensation calorimetry, using the Peltier and Joule effects. Measurements of heats of solution, and of the heats of slow reactions. *J. Chim. Phys.*, **24**, 325.

1932 E. CALVET; Direct thermochemical measurements on organic reactions. The kinetics and heats of saponification of amides. Thèse, Paris.

1933 A. TIAN; *J. Chim. Phys.*, **30**, 665.

1938 (a) E. CALVET; The Adsorption of solvents by soluble substances. The evolution of heat prior to dissolution due to adsorption. *J. Chim. Phys.*, **35**, 69.

(b) Studies on the solubilities of non-electrolytes, and the effect of neutral salts on the solubilities. *J. Chim. Phys.*, **35**, 286.

(c) Studies on solubility and salting-out. *LXXIe Congrès des Sociétés Savantes*, p. 53.

1941 E. CALVET; Calorimetric studies on the adsorption of acetone by nitrocelluloses. *C. R. Acad. Sci., Paris*, **212**, 542; **213**, 216; **214**, 716.

1942 E. CALVET; Vapor pressure of nitrocellulose — acetone gels. *C. R. Acad. Sci., Paris* **214**, 767. On the constitution of nitrocellulose – acetone gels. *Ann. Fac. Sciences de Marseille*, série II, XVI. 1.

E. CALVET and Jeanne COUTELLE; Calorimetric study of the gelatinization of nitrocelluloses by methyl nitrate. *C. R. Acad. Sci., Paris*, **215**, 138.

1943 E. CALVET and A. MAURIZOT; Thermal effects accompanying the adsorption of acetone by the tetranitrate of β-methylglucose and the heptanitrate of β-methylcellobiose. *Ann. Fac. Sci. de Marseille*, série II, XVI, 2.

E. CALVET; *C. R. Acad. Sci., Paris*, **216**, 51.

E. CALVET; Calorimetric study of the gelatinization of nitrocelluloses by esters of acetic acid. *C. R. Acad. Sci., Paris* **217**, 482.

1944 E. CALVET; The heats of adsorption, solvation and solution of very soluble substances: the systems phenol-acetone and diphenylmethane-acetone. *Soc. Française de Physique, Section Méditerranée*, No. 24.

1945 E. CALVET; Specific volumes and heats of dilution of the system cellulose – acetone. *Bull. Soc. Chim. Fr.*, **12**, 5th series, 533.

1946 E. CALVET; The mechanism of solution, from microcalorimetric studies. The gelatinization of nitrocelluloses by a mixture of alcohol–ether. *Mémorial des Services Chimiques de l'Etat*, **32**, 211.

E. CALVET, R. DALBERT and J. CHEDIN; Calorimetric studies on the adsorption of acetone by bakelites. *Mémorial des Services Chimiques de l'Etat*, **32**, 220.

1947 E. CALVET and G. SEBILLE; On the affinity of phenol-acetone mixtures. *Bull. Soc. Chim. Fr.*, 5th series, **14**, 286.

1948 E. CALVET; A new differential microcalorimeter with electrical compensation, *C. R. Acad. Sci., Paris*, **226**, 1702.

1949 E. CALVET, H. THIBON and A. MAILLARD; A thermochemical method for measuring the stability of solutions of sodium aluminate, *C. R. Acad. Sci., Paris*, **228**, 928.

1950 E. CALVET; A microcalorimetric method of measuring the degree of crystallinity of a pure compound. *J. Chim. Analytique Fr.*, **32**, 176.

E. CALVET and J. DHERS-PESSION; The heat of esterification of methanol by nitric acid. *Mémorial des Services Chimiques de l'Etat*, **35**, 45.

E. CALVET, H. THIBON, A. MAILLARD and P. BOIVINET; Solutions of sodium aluminate, and the "decomposition" of these solutions. *Bull. Soc. Chim. Fr.*, **17**, 1308.

1951 E. CALVET and P.-H. HERMANS; Heat of crystallization of cellulose. *J. Polymer Science*, **6**, 33.

E. CALVET, H.THIBON, A. MAILLARD and P. BOIVINET; Microcalorimetric studies on the decomposition of sodium aluminate solutions. *Bull. Soc. Chim. Fr.*, **18**, 402.

E. CALVET; Thermal effects accompanying adsorption. *C. R. Acad. Sci., Paris*, **232**, 964.

E. CALVET; *J. Chim. Phys.*, **48**, 1.

E. CALVET and B. ASTRUC; The adsorption of normal alcohols on hydrargillite: molecular surface areas and heats of adsorption. *Mémorial des Services Chimique de l'Etat*, **36**, 401.

1952 E. CALVET; Thermal effects accompanying adsorption. *J. Polymer Science*, **8**, 163.

1953 E. CALVET; Thermogravimetric studies on adsorption by means of a microcalorimeter with four elements, *C. R. Acad. Sci., Paris*, **236**, 377 and 486.

E. CALVET, P. BOIVINET, M. NOEL, H. THIBON, A. MAILLARD and R. TERTIAN; Studies on alumina gels. *Bull. Soc. Chim. Fr.*, **19**, 99.

E. CALVET and H. THIBON; Studies on alumina gels. *Bull. Soc. Chim. Fr.*, **20**, 696.

1954 E. CALVET and P. BOIVINET; Calibration of a microcalorimeter by means of a radioactive sample emitting constant power output. *C. R. Acad. Sci., Paris*, **238**, 1995.

E. CALVET and J. BROUTY; The measurement of true specific heats: the case of volatile liquids. *C. R. Acad. Sci., Paris*, **238**, 1879.

E. CALVET and J. BROUTY; The measurement of the specific heats of bad conductors using the differential microcalorimeter of Calvet. *C. R. Acad. Sci., Paris*, **239** , 672.

E. CALVET and H. THIBON; The thermokinetics of rehydration of decomposition products of hydrargillite. *C. R. Acad. Sci., Paris*, **239**, 1133.

E. CALVET and P. LONGUET; 25th Congress of Industrial Chemistry, Brussels.

E. CALVET and H. THIBON; The thermokinetics of rehydration of the thermal decomposition products from various trihydrates of alumina. *Bull. Soc. Chim. Fr.*, **278**, 1343.

1955 E. CALVET; *Experimental Thermochemistry*, Vol. 1, Chap. 12, Interscience, New York, 1956.

Max MICHEL; *C. R. Acad. Sci., Paris*, **241**, 1462.

1956 E. CALVET, H. THIBON and J. CHAPUIS-SEITE; Application of microcalorimetric methods to measure the specific surface of the thermal decomposition products of hydrargillite. *Bull. Soc. Chim. Fr.*, **23**, 1939.

1957 E. CALVET, H. THIBON and G. PIALAT; Studies on the thermal decomposition of trihydrates of alumina. *Bull. Soc. Chim. Fr.*, **24**, 225.

Max MICHEL; *C. R. Acad. Sci., Paris*, **244**, 73 and 595.

1958 E. CALVET and F. CAMIA; The derivation of thermogenesis curves from microcalorimeter thermograms. *J. Chim. Phys.*, **55**, 818.

1959 E. CALVET; Recent progress in Microcalorimetry. *J. Chem. Phys. Acad. Sci., U. S. S. R.*, **33**, No. 6, 1161.

E. CALVET and H. THIBON; Studies on the mechanism of the "decomposition" of solutions of sodium aluminate. *Bull. Soc. Chim. Fr.*, **26**, 1577.

1960 E. Calvet and H. Thibon; Application of microcalorimetry to studies on catalysis. *J. Chim. Phys.*, **57**, 5.

E. Calvet, P. Chovin, H. Moureu and H. Tachoire; A micro combustion bomb for use with the Calvet microcalorimeter. *J. Chim. Phys.*, **57**, 593.

1962 E. Calvet; *Experimental Thermochemistry*. Vol. 2, Chap. 17, Interscience—Wiley, New York.

PART 3

BIOLOGICAL APPLICATIONS

by HENRI PRAT

FOREWORD

EACH event taking place in a living system is associated with a production, or absorption of heat. Thus the continuous recording of the thermogenesis *(thermogram)* faithfully detects the smallest fluctuations of the metabolism, and of all vital functions, in animals, plants, bacterial cultures, isolated organs, tissue cultures, etc. The thermograms from such diverse sources enable us to isolate the special characteristics of a given group, to note the changes which are a function of age, and to observe the extent to which individuals of a given group vary from one another. We can also investigate the influence of environment on the thermogenesis, in particular of temperature, humidity, atmospheric composition and of external physical and chemical agents. Thus microcalorimetry offers several experimental procedures which collectively constitute the technique of "microcalorimetric analysis", of wide applicability in the biological sciences, especially in the fields of comparative physiology, ecology, microbiology, agronomy, medicine, pharmacology, and gerontology.

The special characteristics of the Tian-Calvet microcalorimeters make them valuable instruments for biological research. They operate at constant temperature and can be used to provide continuous recordings of the thermogenesis (thermic output) of animals, plants and micro-organisms. Moreover, the calorimeter temperature can be held at any given level between 0° and 50°C, which covers the entire range of all vital phenomena. Studies of thermogenesis can be pursued in several ways, e.g.:

(a) By placing individuals belonging to different species under identical conditions, we may compare their thermograms and deduce specific characteristics useful to physiologists, taxonomists and ecologists.

(b) By recording thermograms of a single individual at different stages of its life, we may follow, as its age increases, the smallest changes occurring in its thermogenesis, hence in its metabolism, in relation to senescence (aging process) thus collecting valuable data for gerontological studies. All the events in the life cycle are faithfully recorded on the thermogram, e.g. metamorphoses and the various circumstances of reproduction.

(c) By placing individuals of the same species, of the same age, under different conditions of environment, nutrition, health, intoxication, etc., we may observe the changes in the thermograms and can thus analyse the effects of the various factors involved: physical, chemical or biotic. Such data are valuable for ecology, dietetics, pathology, pharmacodynamics, toxicology, etc.

(d) In particular we may suppress, by appropriate chemical agents, a given form of functional activity (e.g. muscular, suppressed through anaesthetics). In this way, we can isolate the components of thermogenesis from which the overall thermograms are built up.

To carry out these experiments, the most convenient recording devices are those giving an immediate inscription with pen and ink on paper, such as Sefram "Photodyne", Beckman "Photopen" or Leeds and Northrup "Speedomax" recording microvoltmeters. Photographic recorders are less practical since the results are not observable until the end of the experiment.

In biological research it is often interesting to perform experiments at temperatures below that of the laboratory (between 20° and 0°C). For this purpose the microcalorimeters must be equipped with cooling coils. Also, for biological studies the normal type of calorimetric cell (18 millimeters in diameter, 15 to 20 cc capacity, with 190 thermocouples) is too small to permit the entrance of many organisms (middle sized vertebrates, large seeds or stems, etc.) A more convenient type of apparatus is provided with larger cells (35 mm in diameter, 100 cc in capacity, with 1880 thermocouples) wide enough to contain an adult mouse, a bat, a small bird, fish or lizard. For special studies it is useful to mount the calorimeter on a support allowing rotation about a horizontal axis, so that the inverted cell can receive a growing stem or, if placed horizontally, can hold a mammal in a more natural position than allowed by a vertical cell of the usual type.

As in most other new fields of biological science (genetics, cytology, biochemistry, physiology, ecology, etc) physiological microcalorimetry takes its experimental material from among all groups of living organisms.

In the following chapters we shall briefly describe some applications of this novel technique in relation to:

(I) Plant life, in particular thermogenesis of seed germination and thermogenetic gradients in growing stems.

(II) Development of bacterial cultures, with an analysis of the efficiency of antibiotics, as revealed by the thermograms.

(III) Comparative physiology of various invertebrates, including insects and other arthropods; molluscs, worms, etc.

(IV) Thermogenesis of cold-blooded and warm-blooded vertebrates, with special mention of the effects of anaesthetics, of antipyretics and of the influence of aging.

CHAPTER 13

PLANT THERMOGENESIS

Plant life seems, at first sight, so different from animal life that the profound similarities between the physiological processess of the two kingdoms are frequently overlooked. Yet it is well established that plant protoplasm is free to move within the confines of its cellulose boundaries, that it breathes, and, that when it divides, its chromosomes duplicate themselves in the same way as those of animals. But comparatively little attention has been given to the thermal behaviour of plants, although this provides one of the most precise indications of their functional activity.

1. Thermograms of Germinating Seeds

The heat evolved by seeds during germination is a classical example of plant thermogenesis. Until recently, only crude calorimetric measurements were available, obtained by introducing a thermometer into a mass of moistened seeds placed in a well-insulated container. The progressive temperature rise was noted and, knowing the heat capacity of the mass, the heat evolved was calculated. This method has two serious shortcomings:

(a) It requires a considerable mass of seeds, resulting in asphyxiating conditions in the lower layers.

(b) It involves a progressive increase in temperature during the experiments, which detracts from the physiological value of the results, since it is well known that thermogenesis of living tissues varies considerably with temperature. If the heat flow curve is to have significance, it must be obtained at constant temperature.

With the Tian-Calvet microcalorimeter these difficulties do not arise. First, on account of its sensitivity, it is possible to work with small quantities of seeds, under conditions ensuring satisfactory aeration for each grain. It is even possible, in favorable cases, to work with only a single grain. In addition the small amount of heat produced is rapidly

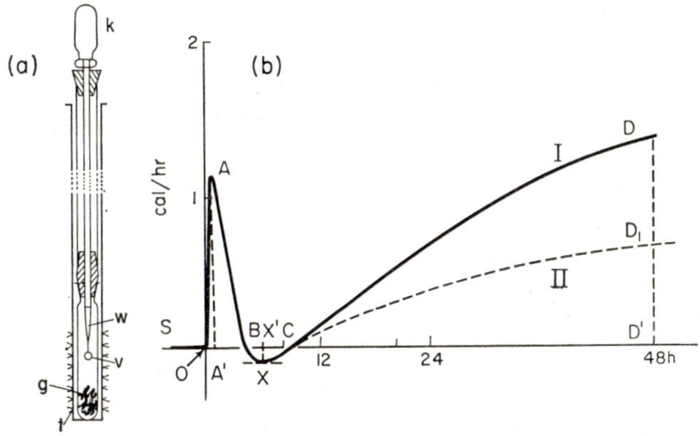

FIG. 13.1(a). Device used for the study of germination. *g*, seeds; *w*, tapered tube containing water; *k*, rubber bulb; *v*, vaseline plug; *t*, thermocouples.

FIG. 13.1 (b). Thermogram obtained with 32 grains of Blackhawk winter wheat, total weight, 1 g, brought into contact with 1 ml of distilled water at 24°C, the experiment being carried out in darkness. O = starting point: moistening of the grains; OAB = physico-chemical thermogenesis; BC = dead time; CD = biological thermogenesis under normal conditions. Curve II shows at CD_1 the thermogenesis of the grains asphyxiated by submersion in 2 ml of water. (Prat and Calvet, 1944)

conducted away into the mass of the calorimeter, so that the temperature of the material does not change significantly during the course of an experiment.

The main device we have used for studying thermogenesis of germination is shown in (a) Fig. 13.1. It consists of a long tapering tube *w*, containing water, provided with a rubber bulb *k*. A plug of Vaseline prevents the water from dripping prematurely. After temperature equilibrium is reached (as shown by the recording of a perfectly horizontal line), the rubber bulb is squeezed and water drops on the grains, the plug of vaseline being washed out by the pressure.

Figure 13.1(b) shows a thermogram obtained by photographic recording from 1 g of Blackhawk winter wheat at 24°C, put in contact with 1 ml of distilled water. This experiment was performed in darkness. At zero time, *O*, water was dropped on the grains; the heat output increased rapidly until after 20 to 40 minutes, it reached a peak of 1.2 calorie-hour. Then the heat flow decreased, and fell to zero after five hours (Fig. 13. 1b, *B*). This first phase *OAB* seems to be merely the result of the hydration of dry seeds; it occurs equally as well with dead grains as with living ones. We call it "*physico-chemical thermogenesis*" to distinguish it from later phases, which involve the activity of living cells. This phase of thermogenesis, which immediately follows the moistening, is particularly important for the life of the plant. It corresponds to the "*hydro-stage*", i.e. the phase during which the plant can only react to moisture.

For a period extending from the fifth to the tenth hour after moistening, the heat evolved is small and, in some cases, there may even be an absorption of heat. We have called this phase "*dead time*" (*BC*, Fig. 13.1b). Then the heat flow rises again gradually (*CD*, Fig. 13.1b), as the physiological processes of germination commence, i.e. increase of the respiratory intensity and intervention of diastases hydrolysing the reserves. We call this phase "*biological thermogenesis*". It does not occur with dead seeds. It corresponds to the beginning of the "*thermo-stage*", i.e. the phase during which the plant reacts to temperature, whilst still remaining sensitive to the presence of water. It is to be noted that, as from the beginning of the thermo-stage, the presence of oxygen, unimportant in the earlier stages, becomes necessary because of the rapid increase of respiration. This may easily be confirmed by subjecting the grains to asphyxiating conditions, e.g. by submersion; the thermogram then rises less rapidly (CD_1, Fig. 13.1b).

2. Influence of Temperature on Thermogenesis

The shape of the thermogram varies as a function of the surrounding temperature. Figure 13.2 shows some recordings of wheat germination at the following temperatures: 17°; 23°.

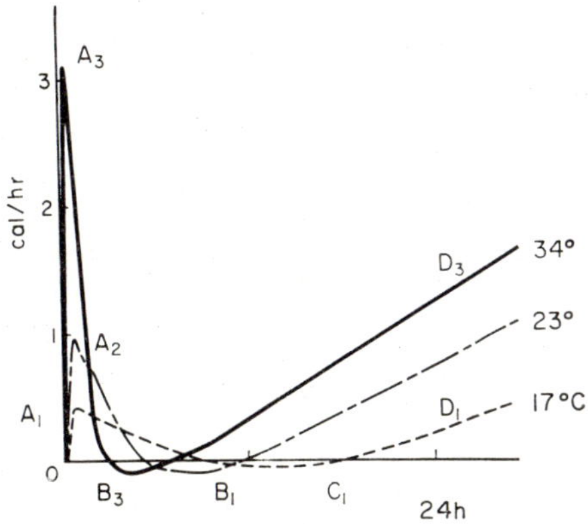

Fig. 13.2. Thermograms of wheat at three different temperatures: (27°, 23° and 33 °C).

34°. Within the limits of these experiments it is seen that a temperature rise:

(a) shortens the duration of the physico-chemical thermogenesis *OB*.

(b) raises the maximum, *A*, of thermogenesis.

(c) shortens the time *OC*, making the biological thermogenesis start earlier and shortening the dead time *BC* (Prat, 1951 and 1952).

These results show that to compare the thermograms of various species, it is necessary to carry out all the experiments at the same temperature.

3. Influence of a Prior Dehydration of the Seeds on the Germination Thermogenesis

The heat released during the physico-chemical thermogenesis is due to the hydration of the seed tissues (stored foodstuffs, protoplasm, cells, walls, etc.) If seeds are subjected to an intense drying before the experiment the amount of heat released after soaking will thus be increased (Prat and Calvet, 1945 a).

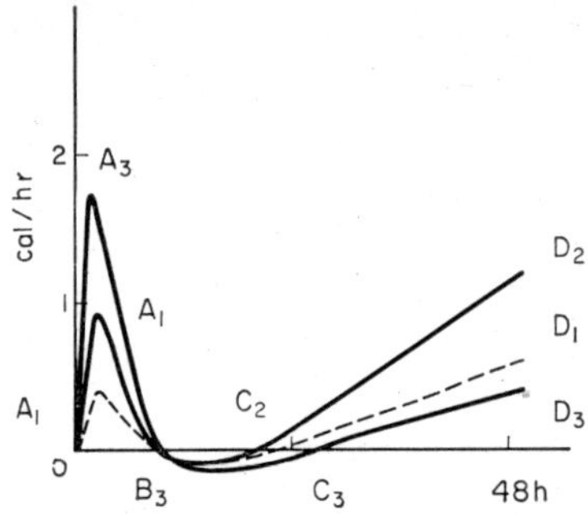

Fig. 13.3. Thermograms of wheat grains previously subjected to vacuum dehydration in the presence of phosphoric acid: for 2 days (OA_2D_2), and 24 days ($OA_3B_3C_3D_3$). Curve OA_1D_1 serves as reference (the seeds were not specially dried before the experiment) (Prat, 1952).

Figure 13.3 shows thermograms obtained with wheat grains dehydrated *in vacuo* in the presence of P_2O_5 for two days (curve OA_2D_2) and 24 days (curve OA_3D_3). Curve OA_1D_1 was obtained using grains not previously dried. It may by seen that the height of the maximum, A, is approximately doubled after two days drying and that, by drying for 24 days, it becomes four times higher that of the control. Since the duration of the physico-chemical thermogenesis is almost unchanged the total amount of heat released during this phase varies approximately in the same proportion as the height, A. The biological thermogenesis may also be increased, (C_2D_2), provided dehydration has not been excessive. Results are representative of a given batch of grains at a given time, but may differ from one experiment to another as conditions vary. To reduce the variables to a minimum one must store the grains in special stoves at constant temperature and constant humidity. Even in this case, variations may occur, as the aging of the seeds continues. Thus the same sample, even when stored under constant conditions, gives results changing progressively from week to week.

In nature it often occurs that a seed, resting on the surface of the soil or buried near the surface, is exposed to considerable variations in humidity. It is of interest to know the effect of these variations on the thermogenesis. When a sample of grains is soaked once, dried and soaked again, then dried and soaked a third time, the duration of the physico-chemical thermogenesis shortens, whilst the maximum A_1 rises; the total sum of heat evolved by the physico-chemical thermogenesis remains nearly constant. It seems that the teguments of the grain become more and more permeable. In addition, the chemical nature of the reserves changes from experiment to experiment, as a result of the enzymatic mobilization caused by each soaking.

The shorter duration of physico-chemical thermogenesis and the increase in its maximum intensity when the seed is subjected to alternate wetting and drying are important because of their ecological consequences. This may be one of the reasons why the seeds of most plants growing in desert climates germinate rapidly as soon as they receive the smallest amount of moisture.

4. Effect of the Aging of Seeds: Dormancy

Modifications in the thermogram of seeds stored under uniform conditions provide a means for study of the dormant state, or dormancy, which is one of the most complex problems of plant physiology, and interesting because of its important applications in agronomy.

It is well known that a seed often cannot germinate immediately after harvesting; it needs first a "post-maturation" under well defined conditions of temperature, humidity and light. This delay is called the "diapause of maturing". In order to terminate this stage, some species need cold, others dryness, others humidity, others darkness; still others need illumination by light of well-defined wave-length and intensity. Furthermore each of these factors must intervene at a specific period in the life of the plant. (P. Chouard, 1951).

This diapause may be related to two different kinds of factors: some of these are accidental, such as a "tegumental inhibition", depending on permeability of the grain shell at each stage; others are intrinsic such as a particular dormancy of the embryo depending only on its aging. Finally, the grain may be kept from germinating through the action of "inhibitors" diffusing from the fruit itself, or through the action of toxins produced by other plants present in the soil. This latter phenomenon is know as *teleintoxication* (G. Bertrand, 1946). We have seen above how the microcalorimeters make it possible to follow changes of the tegumental permeability through alternating states of moisture and dryness; it also allows us to evaluate numerically the dormancy of the grains, and to determine the effects of the various factors, which can either intensify or reduce the dormant state.

We may take as a criterion of germinating activity the value of $DD'48$, (Fig. 13.1b) i.e. the biological thermogenesis observed 48 hr after soaking the seeds. It is high when there is little or no dormancy and, of course, low when the dormant period is long. Three factors should be considered: (a) the age of the grain, (b) its state of preliminary desiccation, (c) the temperature at the moment of hydration. The two former are generally correlated with each other since, under normal storage conditions, the older the seed the more it dehydrates. We have seen in section 3 how drying affects the thermogenesis. The dormancy of cereals is generally *xerolabile*, which means that activation occurs after a few weeks drying. This protects the plant, since the grains are kept from germinating on the ears should the rainy season start before harvesting. Oats display the longest dormant period and ryes the shortest. One may, in general, eliminate these periods either by denudating the grains or by previous dry heating at 35° or 40°C for three or four days. Dehydration increases the "hydrophilic" nature of the seed, as demonstrated by the amplification of the physico-chemical thermogenesis (Fig. 13.3), and thus acts as a kind of "artficial aging". The phenomenon may be compared to a spring which is compressed: the more a grain is dried, the higher the final degree of hydration attained and consequently the larger the amount of heat released.

On the other hand germination may be inhibited by small amounts of soluble substances called "blastokolines", comprising many organic acids: citric, tartaric, malic, caffeic, ferulic, cinnamic, and also alcohols, essential oils, alcaloids, which are often responsible for the dormant state. This leads us to examine more generally the influence exerted on biological thermogenesis by substances dissolved in the water soaking the seeds.

5. Effects of Substances Dissolved in Water

If we add 5% ethyl alcohol to the water used for soaking the seeds, the physico-chemical thermogenesis is not affected, but the biological thermogenesis is delayed, i.e. the dead time is indefinitely prolonged. However the seeds are not killed: when withdrawn from the alcoholic solution and placed into pure water they germinate normally and the biological thermogenesis starts. (Prat, 1944).

Conversely, if we add α-indol acetic acid at 1/10,000,000, i.e. a concentration of 5.7×10^{-7} mol/l, the biological thermogenesis is accelerated, starting after ten hours instead of 15 (at 17°C). After 48 hours, it attains values twice as high as those of controls soaked with pure water. This may be related to the increase of the respiratory intensity observed under the influence of auxins (R. Pratt, 1938). Gibberellic acid also increases thermogenesis of plant tissues, but is less effective on seeds, perhaps on account of their reserves of enzymes, and its action is variable.

Saline solutions also modify the seed thermogenesis: wheat grains soaked with a Sachs medium give a physico-chemical thermogenesis three times higher than with pure water (Brachet, 1949). Iron sulfate at 1/10,000 increases it five times. But, used alone, this salt has an inhibiting effect on the biological thermogenesis; the presence of the other components of the Sachs solution is necessary to neutralize this toxic influence. In contrast to iron, calcium sulfate at 5/10,000 does not modify the physico-chemical phase and speeds up notably the biological thermogenesis. (Prat, 1952).

We have also studied the action of antibiotics on the thermogenesis of germination. At low concentrations there is little effect. Thus one may use these substances for preventing the growth of bacteria during experiments of long duration.

A special mention must be made of the influence exerted by the chlorhydrate of chlorpromazin (Largactil). This drug is used to bring about artificial hibernation for surgical purposes (Laborit 1952). (In Chapter 16, its effects on thermogenesis of mammals will be discussed.) With plants the

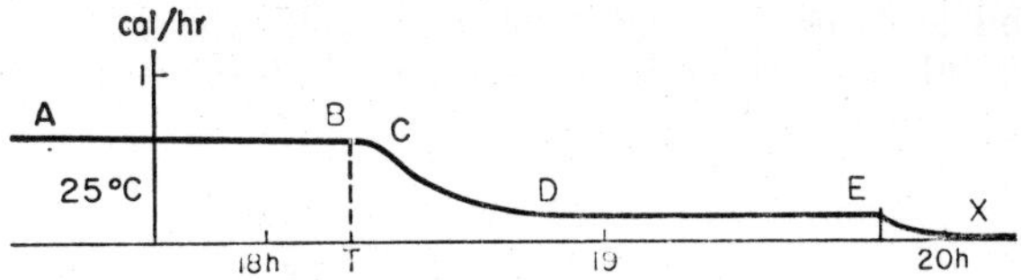

FIG. 13.4. Effects of Largactil on the biological thermogenesis of wheat: 22 grains of Acadia wheat, weighing 1g, wetted with 1 cc water at 25°C. *AB* — thermogenesis at the 18th hour: *BT* — introduce a drop of Largactil solution containing 40 mg; *CDE* — depression of thermogenesis: *EX* — experimental return to zero for control. Note the time scale is here 20 times larger than in Fig. 13.1(b) (Prat, 1957).

following results (Prat, 1957) have been obtained: addition of Largactil in a dose corresponding to a concentration of 1 : 40,000 in the mass provokes a strong drop of the thermogenesis of germinating wheat grains, the thermic output decreasing to 25% of its previous value. This confirms the general narcobiotic influence of chlorpromazin pointed out by Decourt (1954).

6. Specific Characteristics Appearing on Thermograms

In all the thermograms given by seeds of cultivated or wild plants we have found the same fundamental components: physico-chemical thermogenesis, dead time, biological thermogenesis. Yet, the shape of the curve and the numerical values of its elements vary considerably between the species (Prat, 1952). Figure 13.5 compares the curves of thermogenesis of three species: I = celery; II = corn; III = cabbage.

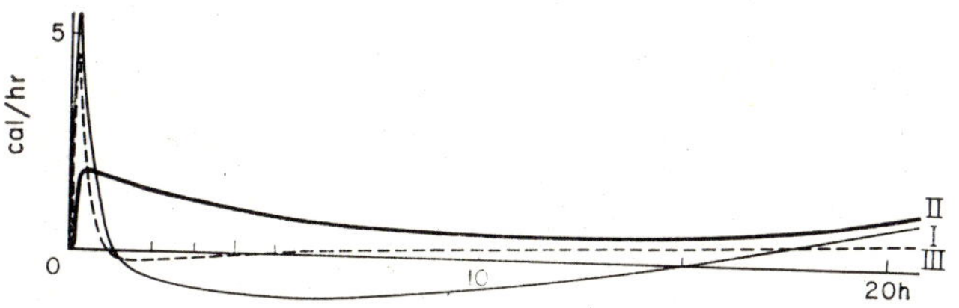

FIG. 13.5. Thermograms of seeds in germination. I—celery (2292 seeds); II — maize (4 seeds); III — (dotted curve) cabbage (236 seeds). Each sample weighed 1 g in the dry state, and was wetted with 1 cc water, at 24°C (Prat, 1952).

As a general rule, the physico-chemical thermogenesis *per gram* of small seeds, has a higher maximum AA' and a shorter time OB than that per gram of large seeds. This is obviously related to their relatively larger surface area per gram, resulting in a higher speed of imbibition. However the relation is far from being strict, and the phenomenon is not governed simply by geometrical factors. For instance carrot seeds give a higher maximum AA' (8 cal/hr/g) than celery seeds (6.4 cal/hr), although they are larger, there being 880 carrot seeds per gram, as opposed to 2292 celery seeds per gram. The same effect occurs with radish seeds, which show a higher AA' (14 cal/h) than cabbage and cress (5 cal/h) although larger (104 seeds per gram against 236 and 552 respectively). It will be noted that the first example involves two Umbelliferеae and the second three Cruciferеae, thus, in both comparisons the seeds are anatomically and histologically similar. This shows that the thermogenesis characteristics are really specific and not the mere result of size ratios.

7. Influence of Ultra-Sonic Vibrations on Germination Thermograms

Busnel and Obolensky (1954) subjected barley seeds to ultra-sonic vibrations and observed an acceleration of their germination partially due to an increase of the permeability of the tegument. With the aid of radio-active phosphorus

P^{32}, they also showed that there is an activation of the metabolism after ultra-sonic treatment.

Obolensky and Calvet have completed this study by recording thermograms of plants after ultra-sonic treatment. They observed that biological thermogenesis is increased, confirming the results obtained with radio-phosphorus.

8. Gradients of Thermogenesis in Plants

A growing plant organ always establishes a system of interdependent gradients, some physiological, some physico-chemical and others which are inscribed in histology and anatomy (see H. Prat, 1945 d, 1948). Among physiological

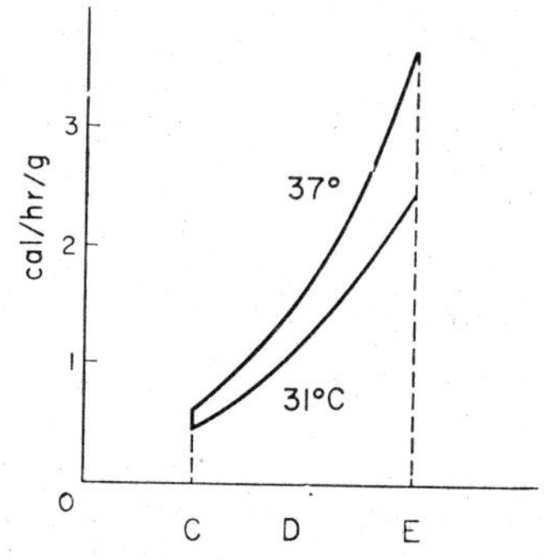

Fig. 13.6. Gradients of thermogenesis in a growing Asparagus stem, at 31° and 37°C. *C* — base of the stem; *D* — half-height; *E* — top.

gradients those of thermogenesis offer a special interest. As stated, plant thermogenesis is chiefly a function of the activity of cell division so that tissues where proliferation is active yield the highest thermal output. Thus measuring gradients of thermogenesis of a growing stem or root corresponds to an evaluation of the mitotical activity along this organ.

By placing previously cut segments from a stem of asparagus in the calorimeter, we have established that there is a thermogenesis gradient from 2.5 cal/hr at the asparagus tip, down to 0.5 cal/hr at the base of the shoot, at 31°C (Fig. 13.6). At 37°C the values are respectively: 0.6 cal/hr at the base and 3.5 at the tip. On the average, the thermic flux appears to be 5 or 6 times higher at the apex than at the base of the shoot when it is actively growing.

Gradients of thermogenesis have also been measured in seedlings. On account of their large size, those of *Persea gratissima* (Avocado pear) constitute a suitable material. At 25° the embryos give an average of 0.25 cal/hr/g for the cotyledons, as opposed to 0.9 cal hr/g for the bud.

Floral organs, chiefly stamen and ovaries, are also the seat of an active thermogenesis at the moment of blossoming. It is well known that an *Arum* inflorescence (Jack in the pulpit), enclosed in its spathe, shows a temperature higher by several degrees than the surrounding air. This may be detected easily with an ordinary thermometer.

For *Tragopogon pratensis* (salsify), the following thermal flows were observed, at 31°: 2 cal/hr/g in the capitule versus 0.07 in the peduncle, i.e. a ratio of 28 to one. For *Dactylis glomerata* (orchard grass) the thermogenesis reaches 0.8 cal/hr/g in the inflorescence, compared to 0.1 in the internodes of the culm.

These facts are connected not with cell-division, but with an exaggeration of the metabolism of the tissues, under the influence of flowering hormones, with, as ultimate result, the death through exhaustion of most of these floral organs.

Conclusion

Thermograms of plants examined above show several characteristics worth specifying before discussing more complex cases:

1. Plant thermogenesis exhibits a great regularity. Its variations are slow and progressive; thermograms of plants do not show the sudden irregularities which occur frequently in the case of animals. The only instance of a steep slope

in plant thermograms occurs in the first phase of germination; but it has been stressed that this is due to a physicochemical, and not to a physiological, phenomenon.

2. Plant thermogenesis is chiefly a result of mitotic activity: the principal seats of heat production are the growing meristems. Nevertheless plant tissues in a state of activity always produce a detectable thermic output, which is an expression of their basal metabolism.

3. The intense thermogenesis displayed by flowers at the time of blossoming is not only due to cell divisions but also to a sudden stimulation of the metabolism. This is a sort of intoxication which causes a rapid increase of respiratory intensity as well as an early differential mortality of the tissues.

4. A rise of thermogenesis in plant tissues is not rigorously proportional to the increase of respiratory intensity, though the two phenomena are generally concomitant. Thus it is not always valid to deduce the heat production from measurements of gas exchanges, as is frequently done. The thermogenesis ought to be directly measured with microcalorimeters.

For studies of plant physiology, thermograms are to be compared in value with electro-cardiograms or electroencephalograms in the study of human physiology. The heat output provides information no less precise than the electric current in respect of the processes occurring in a living organism.

CHAPTER 14

BACTERIAL THERMOGENESIS

Micro-organisms, such as bacteria and yeasts, liberate heat in quantities proportionately large relative to their mass. The thermogenesis from microbes, as from plants, originates primarily in the process of cell-multiplication. The thermograms from bacterial cultures give information on the rate of growth of bacterial colonies, and on the variation with time of the number of bacterial units (Calvet, Fricker and Prat 1945). This information adds materially to that obtainable by use of the nephelometer, and other methods of counting.

1. Thermograms from a Bacterial Culture

Figure 14.1 shows the thermogram resulting from the inoculation of 4 cc of peptone broth with a culture of *B. coli (Escherichia coli)* at 24.4°C. For the first 7 hours the thermogenesis remains quiescent (lag phase), and then increases rapidly from the 10th to the 15th hour (log phase), as the cells divide, increasing their number in geometric progression. The maximum output of 0.32 cal/hr is reached at B_1 after 18 hr. Subsequently, heat output slowly decreases, and the activity of the culture gradually dies away.

It should be noticed that the curve in Fig. 14.1 is not smooth, but shows minute and irregular undulations: this is a general feature of the thermograms given by micro-organisms, which consist of a large number of small physiologically independent units, or colonies. Therefore, the recorded curve is an overall statistical expression, with fluctuations of the sum of these individual metabolisms (Prat, 1953). The case is quite different from that of a plant seed, which has a complex structure of coordinated cells, and gives perfectly regular thermogenesis. Here again, we have an example of specificity in thermograms, and indeed, every category of living being has its own characteristics in this respect.

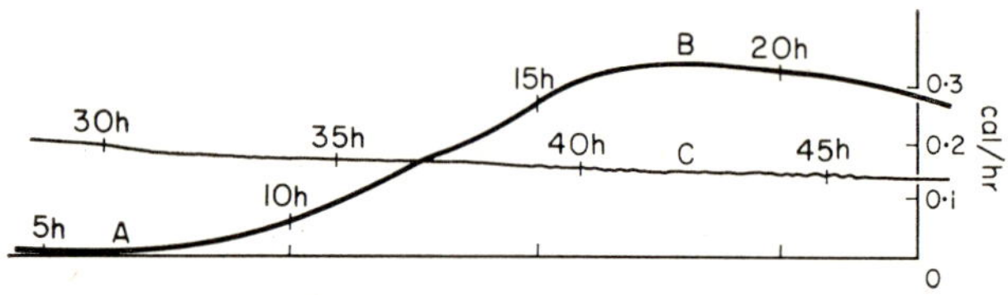

FIG. 14.1. Thermogram of a culture of *B. coli (Escherichia Coli)* in 4 cc of peptone broth at 24.4°. (Calvet and Prat, 1956).

2. EFFECTS OF ANTIBIOTICS

Bacterial growth can be restarted by replenishing the culture with a new supply of broth. Inversely, growth can be retarded by adding an antibiotic substance. Figure 14.2 shows

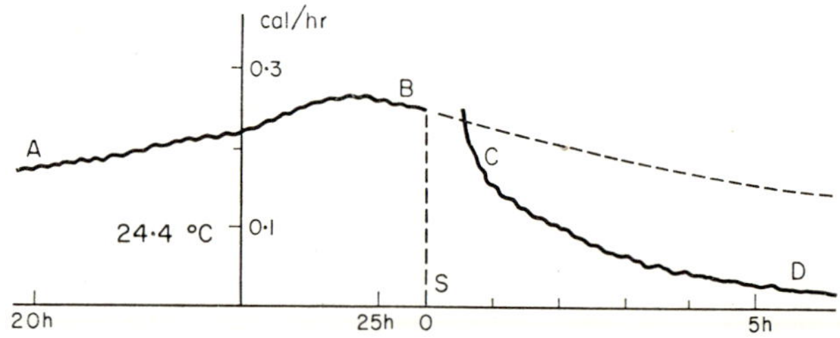

FIG. 14.2.

the effect on a culture of adding, at the point *S*, 1 cc of a solution containing 30 mg of streptomycine. The streptomycine was introduced from a pipette, similar to that shown

(w) in Fig. 13.1: there resulted a marked fall in the heat output, which was reduced to one tenth of its initial value after 6 hr. In this experiment, the calorimeter contents were not stirred, the streptomycine diffusing slowly into the culture. On repeating the experiment with stirring of the contents, the depression of thermogenesis by the streptomycine is very much more rapid.

The microcalorimeter provides us, therefore, with a means of assessing very accurately the activity of antibiotics on a given bacterial family, and should prove valuable in pharmacological laboratories engaged in the assay of the inhibitive power of new antibiotic compounds.

3. Specificity of Bacterial Thermograms

The thermograms given by micro-organisms possess specific characteristics which allow us to identify different bacterial strains, and the issue from any given strain. Certain bacteria, such as the typhoid and paratyphoid bacilli, for example, evolve heat in a fluctuating manner, with occasional outbursts of heat which are "attacks of fever" of the bacteria (Calvet, Fricker and Prat, 1945). The microcalorimeter can thus serve in microbiological laboratories not only to distinguish between different bacterial species, but also to identify the issue from a given species (and identify families which develop rapidly, slowly, at a uniform rate, or by jumps): one can also determine the optimum temperature and the best culture medium for each particular bacterial strain, and measure its appetite for various amino-acids, glucosides, salts, etc.

4. Thermogenesis as a Means of Identification of Resistant Strains

Penicillin-sensitive bacteria show a rapid fall of thermogenesis on adding a drop of penicillin solution to the culture: on the other hand, the issue from these same bacteria may prove to be far less penicillin-sensitive, so that the study of the thermogenesis under these conditions can give a striking demonstration of the emergence of resistant strains.

5. The Evaluation of the Heat Produced by a Single Bacterium

It is of interest to calculate, approximately, the heat output from a single bacterium. We have found, for example, that 1 cc of a *B. coli* culture at the height of cell-multiplication liberates, at 25°C, 0.15 cal/hr, and contains 3×10^8 bacteria. Thus each bacterium produced, on average, 5×10^{-10} cal/hr. We quote this figure merely as an order of magnitude — values we have obtained from other experiments have ranged from 10^{-10} to 2×10^{-9} cal/hr per bacterium, in good agreement with the results obtained by other methods (Bayne-Jones and Rhees, 1929; Winslow and Walker, 1939). These seem small heat outputs in absolute terms, but are however very high relative to the mass of a single bacterium; taking the mass of a single coli bacillus as 5×10^{-2} g, we see that 1 g of bacteria would produce from 100 to 400 cal/hr— i.e. some 25 to 100 times more heat per gram than given by our own bodies.

6. Effect of Bacterial Thermogenesis on the Growth of Infection Centres

The production of heat by bacteria has not generally been taken into account in considering the growth of infection centres. Yet this factor may be far from negligible, for although bacteria are smaller than human cells, they evolve relatively large amounts of heat, and each area of bacterial growth is also an area of excessive thermogenesis. This local heating may be sufficient to kill the neighbouring human cells, or injure them so that they cannot function normally: it is known that the cells of the human internal organs rapidly deteriorate above 40°C.

7. Thermogenesis of Yeasts, and the Microcalorimetric Analysis of Fermentation Processes

Studies of the thermogenesis of yeasts are of considerable interest to the industries making extensive use of fermentation processes, including breweries, bakeries, the wine, cider, vinegar, and cheese industries, etc.

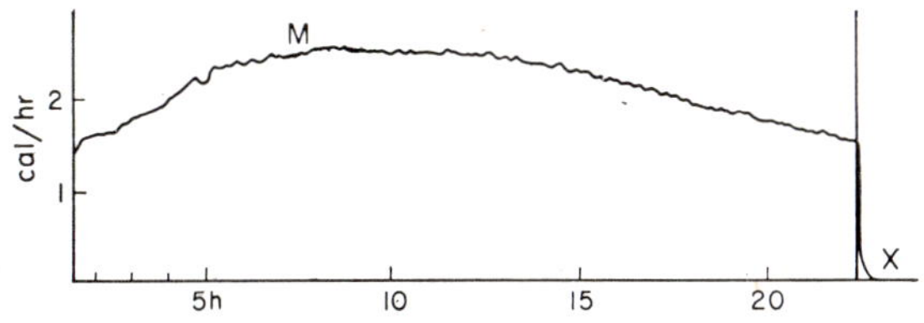

Fig. 14.3. Thermogenesis of a wort of beer during fermentation: 5 cc of wort, inoculated with a decigram of yeast at 31.4°C. (Calvet and Prat, 1956).

Figure 14.3 shows the thermogenesis of a wort of beer during fermentation at 31.4°C. The maximum heat output occurred at *M*, 8 hr after inoculation, and amounted to 2.4 cal/hr per 5 cc. The fluctuations and irregularities in the curve should be noted, and are characteristic of the thermograms from micro-organisms, as mentioned above.

8. Virus and Tissue Cultures

We have not yet studied viruses by microcalorimetric methods, since they cannot be cultivated *in vitro*. However, we hope to by-pass this difficulty through the agency of cultures of animal or plant tissues. The growth of the tissues can be followed in the calorimeter (Calvet and Prat, 1956), and we propose to add a virus at a given stage of tissue growth, and follow the ensuing thermogram: it might then be possible to separate the part of the thermogenesis due to virus growth from that due to the host tissue.

CHAPTER 15

THERMOGENESIS OF ANIMALS OF VARIABLE BODY TEMPERATURE*

The heat produced by animals originates mainly from muscular contractions, and from the glandular activities controlled by the nervous system; on the contrary the main thermogenesis in bacteria and plants is due to cell-division. These differences in the origin of thermogenesis reflect themselves in the respective thermograms: those given by animals are less smooth and regular than those given by plants, and are subject to larger and more rapid variations than those given by micro-organisms.

1. Thermogenesis of an Insect

Figure 15.1 represents a thermogram given by a cockroach *(Periplaneta americana)*. We might note, firstly, the linear parts of the curve, such as C, where the heat output is uniform (or only slightly fluctuating) and small in magnitude, of the order 0.5 cal/hr/g at 25°C. These are "economy periods", lasting for an hour or more, during which the animal is resting, immobile, and its respiratory activity is at a very low ebb (Prat, 1954, 1955).

During the economy periods (periods of apnoea, Chauvin 1949; Wigglesworth, 1939) the spiracles of the cockroach are closed but, from time to time, they open without any

*Poikilotherms, or "cold-blooded" animals.

external stimulation, and the insect enters a phase of active thermogenesis: its thermal output then rises within a few minutes to 5–6 cal/hr/g (A). These spontaneous paroxysms are generally accompanied by a number of secondary maxima but, after 1/2 – 1 hr., the thermal output falls back and there begins another period of rest. During a paroxysm, the insect behaves in an agitated manner, but it would be incorrect to conclude that the increased thermogenesis is solely due to muscular contractions. The phenomenon is

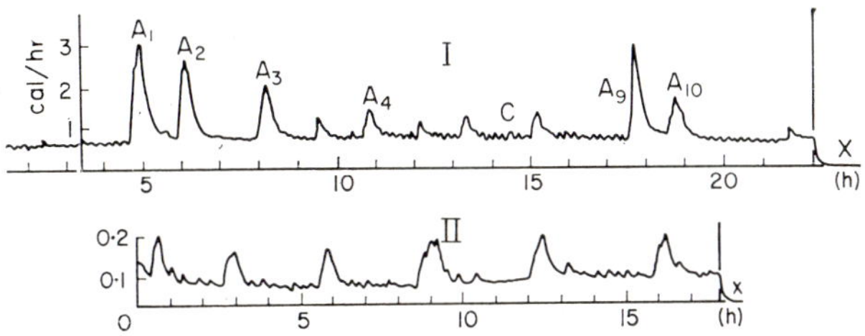

FIG. 15.1. Thermograms of cockroaches *(Periplaneta americana)*. (i) — Adult male, weighing 0.731 g at 24.9°C. A_1–A_{10}, spontaneous paroxyms: C – economy region: X — zero marker. (ii) — Nymph after the first skin-shedding, weighing 0.041 g at 31.4°C (Prat, 1954).

much more complex, and probably the entire metabolism of the tissues is affected. It is to be noted that, during an active phase, the spiracles open and respiratory activity becomes marked.

During an anaerobic rest-period, the insect avoids asphyxia by accumulating lactic acid (Slater, 1927, 1928), thus contracting a "debt of oxygen" which must be restored during the highly exothermic phases of active respiration.

Most insects have the ability to live under conditions of economy as well as of activity, a property we have described as 'thermogenetic elasticity', the measure of this elasticity being the ratio M/m of the maximum and minimum heat outputs. This ratio varies from 2 to 6 for an adult cockroach in the temperature range 24 to 37°C (taking only spontaneous paroxysms into account), and can be increased by inducing paroxysms by physical or chemical provocation.

A large thermogenetic elasticity is a considerable advantage to an animal in the struggle for existence, the long rest periods enabling it to conserve energy and available oxygen, and to meet the need, when necessary, for extremely intense effort. The relatively high thermogenetic elasticity of cockroaches enables us to understand the hardiness of these insects, which can live without both food and air for long periods, without losing their agility.

2. Variations of Thermogenesis during the Metamorphoses of an Insect

The shape of the thermograms given by an insect varies with its age and state of development. Figure 15.1 (II) shows the thermogenesis of a young cockroach nymph, after shedding its first skin. The paroxysms are less intense, and the economy level is higher and less regular than in the case of an adult: correspondingly, the thermogenetic elasticity is smaller.

The changes in the thermogram of an insect undergoing whole metamorphosis are much more striking. Figure 15.2 (I) shows the curve AB given by a caterpillar of *Galleria mellonella*. The curve is irregular, with several small maxima and minima. The trace CD, in Fig. 15.2 (II), which is very uniform, followed the transformation of the caterpillar into a pupa. The considerable perturbation, EFG, coincided with the emergence of the butterfly. Over GH, the butterfly, very wet, and exhausted by the effort of discarding its chrysalis shell, gave first a low steady thermogenesis, until at H', after 4 hr, there occurred a slight paroxysm. Finally Fig. 15.2 (III) shows the thermogram of a male adult butterfly, one day after its emergence from the chrysalis. Here we see a complicated paroxysm, JKL, lasting 5 hr, interspersed between two periods of low thermogenesis.

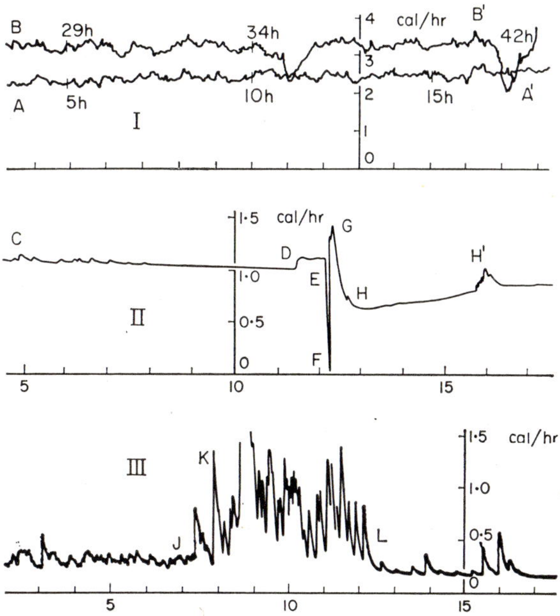

FIG. 15.2. Changes in the thermogram of a butterfly *(Galleria mellonella)* during its metamorphoses. (I) — Caterpillar, shortly before entering the pupa stage, at 24.9°C. (II) — *CD* — chrysalis, weighing 0.19 g at 31.4°C; *EFGH* – emergence of the butterfly. *H'*: small spontaneous paroxysm (III) — Male butterfly weighing 0.06 g at 24°C; *JKL*: Strong spontaneous paroxysm (Prat, 1954).

3. Specific Characteristics of Insect Thermograms

The thermograms given by adult insects, or by their larvae in comparable states of development, show definite characteristics of the species: compare, e.g., the curves 15.1 (I) and 15.2 (III), given by an adult cockroach and an adult butterfly respectively. We have recorded the thermograms given by several species of orthoptera, coleoptera, dictyoptera, lepidoptera, diptera, etc., and can recognize quite clearly characteristics of the curves given by these different orders.

By using the microcalorimeter at maximum sensitivity, it is possible to trace the thermograms of insects of very small size. Figure 15.3 was given by a vinegar-fly weighing only 1 mg. At 24.9°C, it gave a heat output of 0.03 cal/hr, which is a quite high value relative to mass, i.e. 30 cal/hr/g (Prat, 1954).

In contrast to the cockroach, the level parts of the curve here correspond to maximum activity, punctuated at hourly or half-hourly intervals by short sharp falls to inactivity. In general, the thermogenesis by diptera is more active than by dictyoptera. Figure 15.3 shows clearly the point of

FIG. 15.3. Thermogram of a small female vinegar-fly *(Drosophilia melanogaster)*, weighing 1 mg at 24.9°C M – death. (Prat, 1954).

death of the fly at *M*, and the series of depressions just before death, occurring at a more frequent rate than normally.

It is known that mutations in poikilotherms become more frequent at high temperatures. The fact that the number of mutations is proportionally greater for diptera (especially so in drosophilia) than for dictyoptera is possibly related to the stronger thermogenesis shown by the former (compare Figs. 15.3, 15.1).

4. MODIFICATIONS OF INSECT THERMOGENESIS BY PHYSICAL AND CHEMICAL AGENTS

The influence of external physical and chemical agents on insect thermogenesis can be clearly demonstrated by microcalorimetric studies. Figure 15.4, for instance, records the effect of alcohol vapour on the thermogenesis of a cockroach (Prat, 1956). Only certain adult males, in a particularly receptive state, react in this way. The first whiff of alcohol vapour caused the maximum Z_1B_1, which represents a more powerful thermogenesis than given by the spontaneous paroxysms A_2, A_3, A_4. A second dose of alcohol vapour, given 3 hr later, induced the response Z_2B_2, stronger and more prolonged than Z_1B_1. Finally, a third dose, given 16 hr later, induced the very intense paroxysm Z_3B_3C, which

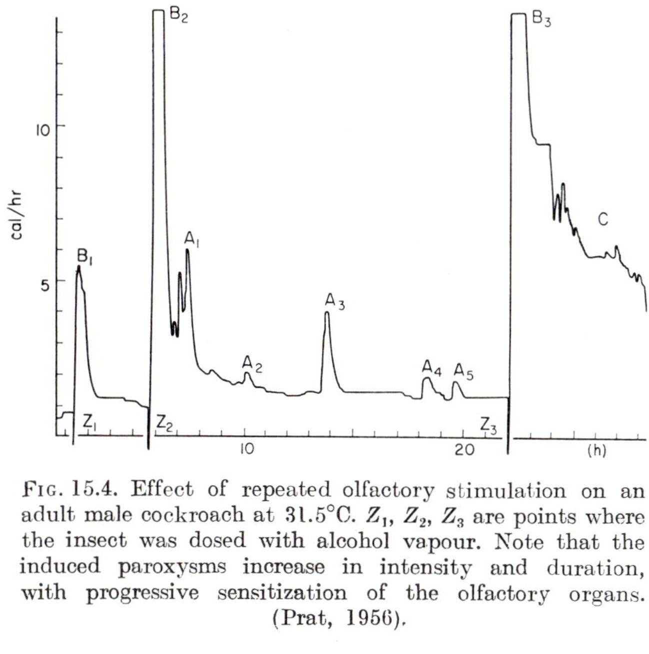

FIG. 15.4. Effect of repeated olfactory stimulation on an adult male cockroach at 31.5°C. Z_1, Z_2, Z_3 are points where the insect was dosed with alcohol vapour. Note that the induced paroxysms increase in intensity and duration, with progressive sensitization of the olfactory organs. (Prat, 1956).

lasted for more than 15 hr without returning to the economy level, and ended in the complete exhaustion of the animal. However, if the antennae of the insect are cut off (thus removing the olfactory organs), it becomes completely indifferent to alcohol vapours: the effect is therefore not due to intoxication, but to a progressive sensitization through the olfactory system.

Apart from the value of studies of this type in respect of the general physiology of insects, they are of considerable practical use in the quantitative measurement of the efficiency of insecticides.

During a certain phase of adult life, the male cockroach reacts to the odour diffused by the female, the thermogenesis from this stimulus closely resembling that due to alcohol vapour (Prat, 1956). We have also shown that a decapitated cockroach continues to survive for more than a week, and that the head itself will live for more than 24 hr; both parts give thermograms different from that of the complete insect, but showing the same feature of spontaneous maxima.

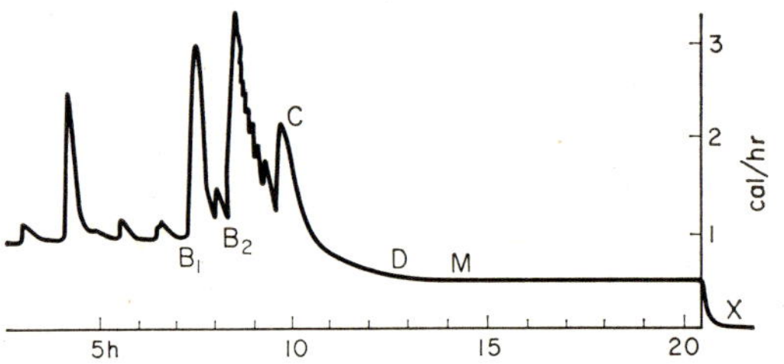

Fig. 15.5. Effect of asphyxia on the thermogenesis of an insect. Nymph of *Periplaneta* after the 6th shedding, weight 0.54 g at 31.4°C, in a sealed cell. B_1C — paroxysms due to asphyxia, after 8 hr confinement; D — coma; M — death; X — zero check (Prat, 1954).

If an insect is placed in a sealed microcalorimetric cell, in the absence of alkali, the thermogram begins to become irregular when the CO_2 concentration reaches a certain level. This is shown by Fig. 15.5, given by a cockroach nymph at 37.4°C: the period of asphyxia, lasting 3 hr, is marked by the series of paroxysms B_1B_2C, followed by a depression of thermogenesis, and eventual death. We might note that some residual thermogenesis continues in the dead body.

It is easily shown that it is the accumulation of CO_2 and not lack of oxygen which causes asphyxia, since the animal will live for several days, exhibiting normal thermogenesis, under the same experimental conditions, if a piece of cottonwool saturated with a solution of caustic soda is included in the sealed cell.

5. Thermograms of Myriapods and other Invertebrates

We have recorded the thermograms of numerous invertebrates, including various groups of insects, myriapods, crustucea, molluscs, and worms. An example is shown in Fig. 15.6, given by a myriapod, *lithobius*. These creatures

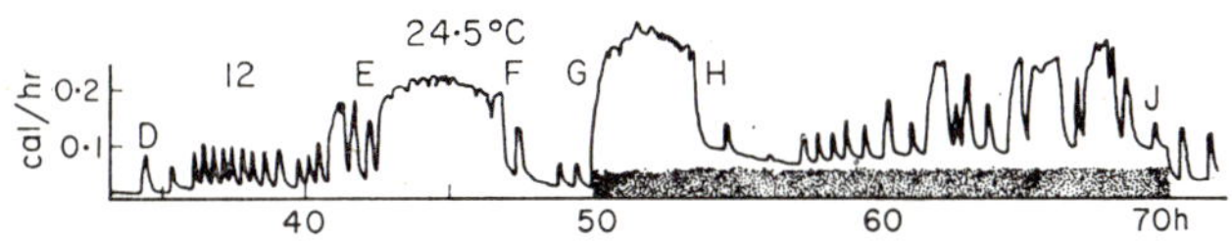

Fig. 15.6. Thermogram of a myriapod *(lithobius)*, weighing 0.125 g at 24.5°C. The dotted area covers the period of illumination of the cell (Calvet and Prat, 1956).

can live for a week in the calorimeter without food. After 50 hr, the calorimeter cell was illuminated (shaded part of thermogram): the immediate effect of the light was the paroxysm *GH* lasting for 4 hr.

Worms are also suitable subjects for the study of thermogenesis gradients in animals.

6. Thermograms of Vertebrates of Variable Temperature: Reptiles and Amphibians

The so-called "cold blooded" vertebrates, which include fishes, amphibians and reptiles, are of special interest among poikilotherms, as they compare in many respects, anatomically and physiologically, with homiotherms. Figure 15.7 (I) shows the thermogram of a snake (Storeria): as in the case of the cockroach, the thermogram has flat sections of uniform, low heat output, corresponding to economy regions, alternating with phases of paroxysm. Snakes are well-known to be able to live for long periods without food — a faculty related to the long periods of living at an economy-level, during which the animal conserves its energy.

The thermograms of amphibians, which include frogs, salamanders, and tritons, have slightly different characteristics, but show the same feature of occasional breaks in thermogenetic activity (Fig. 15.7 (II)).

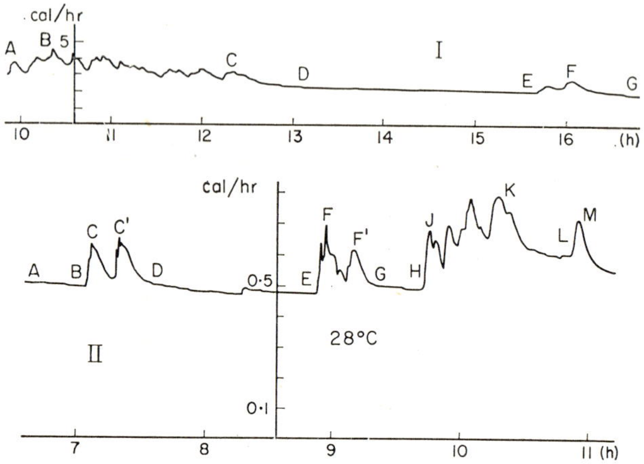

FIG. 15.7(I). Thermogram of a snake *(storeria occipitomaculata)*, male, wt. 2.85 g, at 28°C. *DE* — economy region. (II). Thermogram of a batracia *(desmognathus fuscus)*, wt. 1.05 g at 28°C. *AB*, *DE* — economy regions (Prat, 1957).

CHAPTER 16

THERMOGENESIS OF HOMEOTHERMS

(ANIMALS OF CONSTANT BODY TEMPERATURE)

For the study of vertebrates it is necessary to use the new Calvet micro-calorimeter provided with large cells (35 mm in diameter) and 1880 thermocouples. With the usual small size cells (18 mm in diameter) the only vertebrates suitable for experiments are small snakes, lizards, young amphibians and, alone among mammals, new-born mice. But, with the larger size cells, experiments on adult mice, frogs, bats, small birds, fishes, etc., become possible.

Among the results obtained so far are the recording of changes in the thermogram of a mouse as a result of senescence and, at each age, the variations of the thermic output provoked by various chemical agents: anaesthetics, curarizing, spasmolytics, poisons, etc. Comparison of thermograms of homeotherms (mice) and poikilotherms (snakes, lizards, amphibians) leads to useful observations on the types of metabolism in these groups.

1. Thermograms of Young Mice

New-born mice can be studied either in small or in large size micro-calorimeters. These animals are well-suited to calorimetric study, being able to withstand several days of seclusion in the calorimetric cell without food, and to endure a wide range of temperature. In all these experiments we used only white mice, raised in our laboratories *(Mus musculus albinos)*.

Figure 16.1 shows the thermograms given by new-born mice, one-day old, at temperatures of 25.1°C and 37.1°C respectively. In the first case (*AB*), the mean heat output was 2.7 cal/hr/g; in the second (*CD*), 7.5 cal/hr/g — i.e. 2.7 times larger.

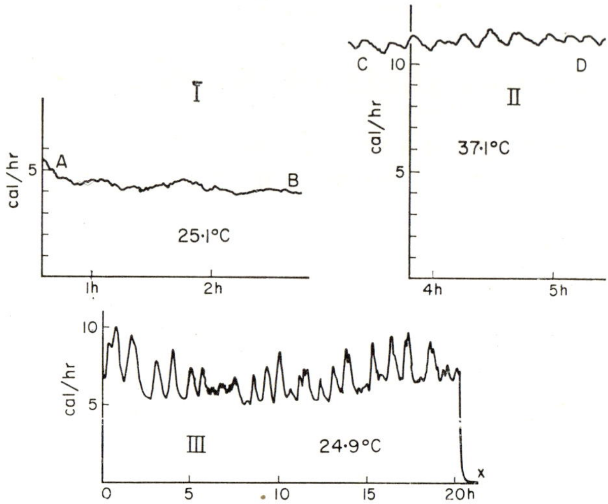

FIG. 16.1. Thermograms of young mice. (I) — Mouse, 1 day old, wt. 1.45 g at 25.1°C. (II) — Mouse, 1 day old, wt. 1.55 g at 37.1°C. (III) — Mouse, 8 days old, w. 1.60 g at 24.9° C. (Prat, 1957).

Thus it seems that the new-born mouse has not yet acquired the ability of regulating its thermal output, the thermogenesis decreasing with a fall in temperature. The thermograms are, nevertheless quite distinct from those of poikilotherms, and there is no evidence of the 'economy regions' characteristic of the latter; on the contrary, there is continuous heat output, sometimes irregular as in *AB*, sometimes with more or less regular fluctuations, as in *CD*.

The thermogram in 16.1 (III), given by a mouse 8 days old, shows much more pronounced undulations in heat output, and indicates that the animal is beginning to acquire the ability of thermostatic control. As in the case of the butterfly (Fig. 15.2), the thermograms undergo changes in form with age of the individual.

From comparisons of thermograms given by numerous young mice, the following conclusions can be drawn:

1. Individual differences in weight are considerable, even between animals belonging to the same litter: at the age of one day, the weight of mice varies from 1.19 to 1.86 g; at 2 days, from 1.32 to 2.19, i. e. a variation of 67% from the minimum; at 7 days the weight ranges from 1.60 to 2.64 g.

2. Individual variations of relative (as well as of absolute) thermogenesis are also very high; after one day, the minima *(m)* range from 3.3 to 6.5 cal/hr/g; the maxima (*M*), from 4.2 to 7.3. From the beginning of the experiment, some animals give a decreasing thermogenesis whereas others maintain it on a constant level.

3. There is no definite relationship between the relative thermogenesis and the weight: a large individual may well have a higher relative thermogenesis than a small one—which is contrary to the general view—or inversely. Besides, even a superficial examination of the animals shows that some are very active, while others are almost inert; this occurs within the same litter, and is not at all related to their sizes. Large animals may be slow and lazy, and small ones very active, or inversely. Individual differences in their metabolisms are, hence, considerable, and it appears in their thermogenesis.

2. Effects of Aging on Thermogenesis

The shape and intensity of the thermograms vary widely as the mouse ages. Figure 16.2 shows these variations from birth to 30 days. The young mouse two or three hours after birth, displays only a faint and slightly undulating thermic flow. One day later the thermogenesis becomes more intense and more irregular. The other thermograms, taken respectively at 3, 7, 12, 18 and 30 days, show a progressive elevation of the absolute thermogenesis and also an enlargement of the irregularities with many peaks and depressions.

Figure 16.3 presents the variations of the average thermogenesis taken on more than 400 mice of all ages from birth

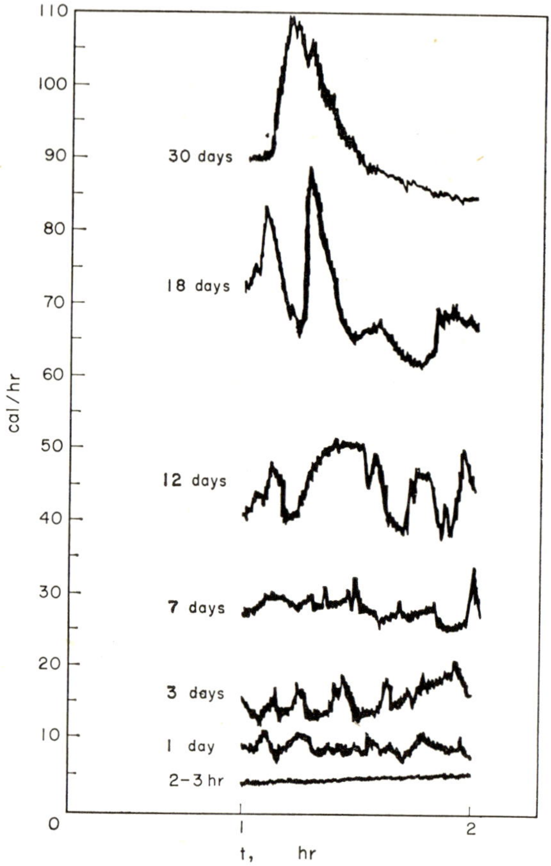

FIG. 16.2. Thermograms of a mouse taken respectively at various ages from birth to 30 days: temperature 23°—27°C. Hours in abscissae are counted from the beginning of each experiment.

to 80 days. Curve *A*, giving the absolute thermic flow, shows that the production of heat increases rapidly during the first month, but more slowly afterwards. Curve *R* gives

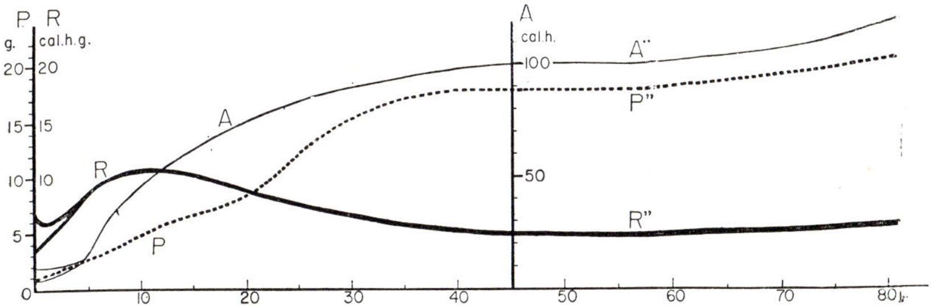

FIG. 16.3. Variations of the average absolute thermogenesis (*A*), weight (*P*) and relative thermogenesis (*R*) as a function of aging of mice from birth to 80 days. The double „roots" of curves *R* and *A* are given respectively by two series of experiments at different temperatures, the upper one between 30 and 37°C., the lower between 19 and 26°C. After the 5th day (homeothermy being reached), the two curves merge.

the relative thermogenesis, i.e., the thermic flow per gram of animal. It begins to grow rapidly until the 10th day; decreases until the 50th day and increases again, but very slowly, after the 60th day. Curve *P* gives the variations of the average weight. It increases rapidly until the 40th day, remains almost stationary from the 40th until the 60th day, then increases again slowly. Male animals have an average weight higher than female.

3. Effects of the Surrounding Temperature on Thermogenesis

Young mice, during the first week of their life, are not yet homeotherms. Figure 16.4 shows thermograms given by two individuals both one day old and of similar weight (1.45 and 1.55 g), kept respectively at 25.1° and 37.1°. It is clear that, at this age, the subjects behave as poikilotherms: at low temperature, their thermogenesis is low and floating; at a higher temperature, it is higher and steadier, with more

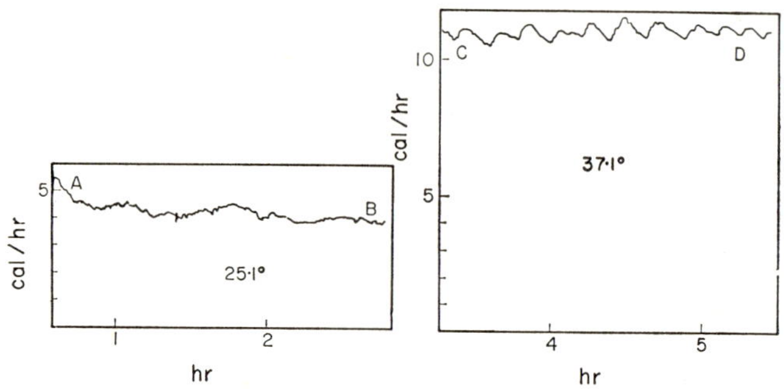

FIG. 16.4. Normal thermograms of a one day old mouse: *AB*, at 25.1°C (1.45 g); *BC*, at 37.1° (1.55 g).

regular undulations. In this case, a rise of 12°C produces an increase of thermogenesis from 4 to 11 cal/hr, i. e. of 175%. This result is sufficient to prove that the young mouse is not yet a homeotherm. However, the case is not exactly similar to poikilothermy. To fulfil the later condition it is necessary to give the subject appropriate chemicals (see section 6) in order to introduce in its thermogram the maintenance level plateaux characteristic of cold blooded animals.

If a young mouse is maintained for several hours at a temperature over 37° great disturbances result in its thermogram. Figure 16.5 shows a result obtained at 37.5°C. In order to avoid asphyxiation, a cell closed only with a cotton pad and containing a small quantity of sodium hydroxide was used. The optimum breathing conditions are thus reached: carbon dioxide was removed and oxygen diffused through the cotton pad. Such precautions are necessary at this relatively high temperature, when active consumption of oxygen occurs.

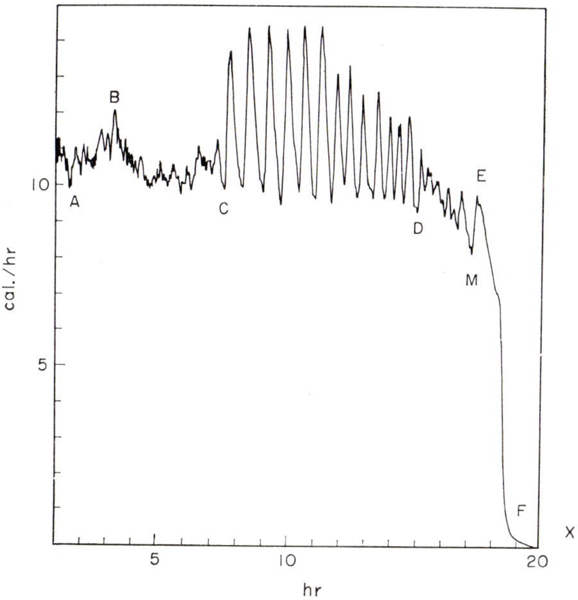

FIG. 16.5. Young mouse, 24 hr old, weighing 1.29 g, at 37.5°C; cell closed with a cotton pad and containing sodium hydroxide (combination of devices *b* and *e* of Fig. 13.1). *CD:* large thermal oscillations due to too high a surrounding temperature. *DE*, phase of the *pre-mortem* irregularities. M = death; X = verification of zero.

During the first period of the experiment the heat flow, (already high from the start: 8.4 cal/hr/g), continued to rise until *B*. This progressive thermogenesis is quite different from that observed at lower temperatures (when the curve starts with a depression). Here, the decline occurred only after the third hour (*BC*).

After seven hours, large oscillations, *CD* occurred, of a type not seen previously. Their amplitude reached 3 to 4 cal/hr/g, with a ratio M/m = 15.7. Six of these large undulations occurred within 4 hr, each having a period of 40 min and showing a fairly constant amplitude. In the following 3 hr, there were 7 further oscillations, but with decreasing amplitudes and shorter periods (about 25 min.)

Finally, after the 15th hour (D), these large oscillations disappeared, and were replaced by small irregular fluctuations of the usual type, with a decreasing flow DE. At the 18th hour a final, *pre mortem* paroxysm occurred, followed by a depression EM leading to death (M) at the 19th hour. Then, over a period of an hour, the zero was reached (FX), with no residual thermogenesis from the dead body.

If these results are compared with those obtained between 20 and 32°C, it may be noticed that the temperature of 37.5°, obviously too high, causes the premature death of the new-born mouse, after a phase of disturbance of the thermogenesis manifested by wide and regular oscillations (CD in Fig. 16.5). These maxima reach 11.5 cal/hr/g of animal, a very high value relative to all those obtained below 32°C (3 to 7 cal/hr/g for animals of the same age). The difference between the highest and the lowest flows may reach 4 cal/hr/g (instead of 1.8), and the ratio M/m may equal 1.5 (instead of 1.1 to 1.4).

Oscillations CD, of Fig. 16.5, represent a physiological reaction of a healthy animal to a surrounding temperature which is too high, i.e. to a special kind of stress. We will call them "hyperthermal oscillations".

4. Effects of Anaesthesia on Thermogenesis

All thermograms of new-born mice presented so far have shown continual oscillations, sometimes regular, sometimes irregular, sometimes small, sometimes large, but always present. In order to discover their cause, experiments have been carried out on animals under the influence of anaesthetics (H. Prat, 1955). All fluctuations of the heat flow due to variations in the muscular activity were thus removed. Various anaesthetics have been tried: chloroform and ether, applied as vapors, Nembutal in injections, and Nembutal administered bucally. The best results have been obtained with the latter method. In fact, with the first two procedures, it proved rather difficult to control the anaesthesia, either because it was insufficient, (the animal remaining active) or too strong, resulting in asphyxiation and death. The following recordings

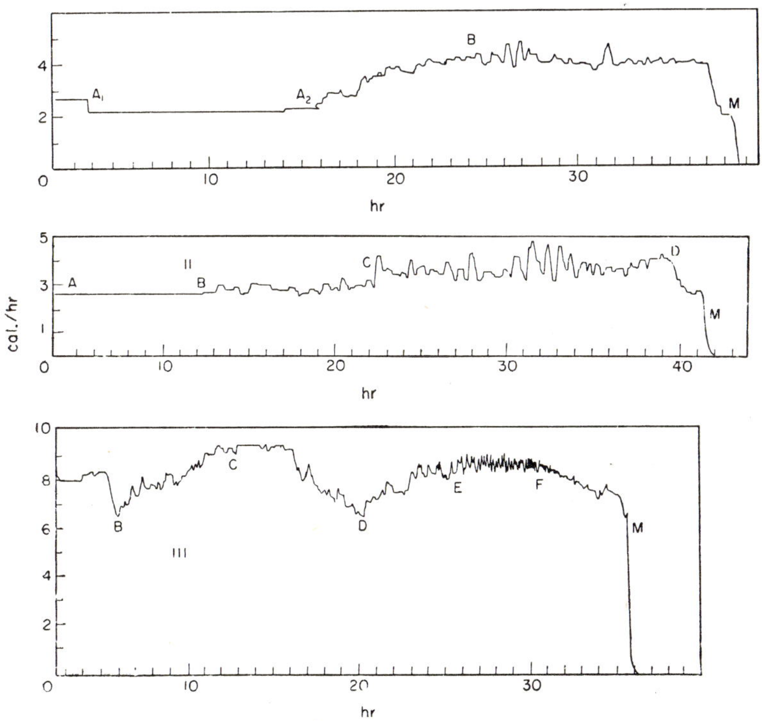

FIG. 16.6. Thermograms of anaesthetized young mice (1 mg of Nembutal, in a 2% solution administered buccally at 32°C. Ages and weight: I, 24 hr, 1.25 g; II, 12 hr, 1.05 g; III, 48 hr; 1.42 g.

have been obtained under optimum breathing conditions: (32°C, in the presence of sodium hydroxide, the cell being closed with a cotton pad to allow the passage of oxygen).

In Fig. 16.6, curve I is the thermogram of a 24 hours old mouse weighing 1·25 g, having taken at the beginning of the experiment one drop of a 2% Nembutal solution (i.e. 1 mg of Nembutal). From A_1 to A_2, (that is for 16 hr), anaesthesia was complete, and a perfectly regular thermogenesis was recorded, with a constant flow of 2 cal/hr, i.e. 1.6 cal/hr/g of animal. After 16 hr, the mouse awoke as shown by an immediate rise of thermogenesis (A_2B), reaching more than 4 cal/hr, and by the reappearance of the usual irregularities. This state was maintained for 21 hr; thermogenesis declined in the 37th hour, death occurring two hours later.

Thermogram II of Fig. 16.6 is that of a very young animal (12 hours old at the beginning of the experiment) weighing only 1.05 g, and having received by mouth, one drop of a 2% Nembutal solution. At the beginning of the experiment a perfect regularity of the thermogenesis (*AB*) is again observed, showing a complete anaesthesia, with a flow of about 2.5 cal/hr. This state lasts 13 hours; then, from the 14th to the 23th hour (*BC*), the animal develops a semi-activity, with a slight rise of the thermogenesis and a reappearance of still weak irregularities. After 23 hr (*C*), the thermogenesis rises sharply, reaching 4 cal/hr, and the fluctuations become sharper too, showing the return to normal behaviour. This state lasts till the 39th hour; then, the thermogenesis starts dropping at *D* until the animal dies (*M*), after 41 hr of imprisonment (a normal life-time under these conditions).

Curves I and II of Fig. 16.6 from A_1 A_2 and *AB*, provide good examples of total anaesthesia as recorded by microcalorimetry. The fact that the animals wake up and regain a normal state of activity for 21 and 27 hr respectively demonstrates that this phenomenon is not a pre-agonic one, but only a temporary anaesthesia without toxic effects. Should the experiments cease, say, after the 30th hour, the subjects would have pursued life normally.

The same amount of the soporific is not sufficient to produce complete anaesthesia in an older animal. Curve III of Fig. 16.16 is the thermogram of a 48 hr old mouse weighing 1.42 g, having received one drop of a 2% Nembutal solution. Only after 5 hr (*B*) is there a decline of thermogenesis, followed by a rise (*BC*). After 18 hr another depression appears (*CD*), but, here again, the sleep is not deep. A final rise appears in *EF*, and death followed, after 36 hr. Rapid and small oscillations of the heat flow appear between the 26th 31st hours, with an average period of 10 min and amplitude of 0.8 cal/hr. Curve III shows, that, even if anaesthesia is not complete, thermogenesis is altered by the intervention of the hypnogeneous drug. This is manifested through two effects: (1) appearance of phases of declining heat flow, separated by temporary rises; (2) changes in the usual small, arythmic undulations, their amplitude becoming now smaller, now larger. Such anomalies are observed either when the quantity of drug is insufficient to produce a complete anaesthesia or in the phase of partial anaesthesia following the awakening (half awakenness). Both cases produce the same condition, i.e. a sub-limit concentration of the anaesthetic in the blood (in the second case, a part of the drug has been eliminated).

To obtain a complete anaesthesia with animals more than a day old, the quantity of soporific has to be increased progressively, by at least as many drops of the 2% Nembutal solution (i.e. milligrams) as the animal is days old (a purely empirical rule). For some of them it was even necessary to exceed this dose. It may be noticed that this increase is greater than the increase of weight of the subject during the same period. Hence the resistance of the young mouse to the soporific is increasing: in the first day one mg per gram was sufficient to produce complete regularity in the thermogenesis, whereas, on the 8th day, 3 mg per gram were necessary to have a comparable effect. A new application of microcalorimetry appears here, i.e. a way to measure the degree of activity of a soporific, and also to compare the sensitivity of various animals to the same agent, according to age, heredity, nutrition or pathological state.

Muscular activity seems to be the principal cause of the fluctuations of animal thermogenesis in a state of awakeness. When the activity is suppressed by anaesthesia, the heat flow becomes perfectly uniform, expressing only the lowest basal metabolism. Yet, it would be erroneous to relate all the fluctuations observed to differences in muscular activity only. Anaesthesia not only modifies the muscular activity but also other forms of physiological functions: secretions, cellular metabolism, circulation, etc.—all of which must be taken in account to explain the thorough regularization of the thermic output of the anaesthetized animal.

These statements may be linked to those made in a previous chapter on Insects, Myriapods and Crustacea. When a Coleoptera "simulates death", when a millipede coils up into a spiral or a woodlouse into a ball, their thermograms become linear, just as that of an anaesthetized mouse. Thus, from the thermogenetic point of view, it is as if the former animals have anaesthetized themselves voluntarily, retaining only a slow and uniform thermogenesis, permitting an extreme economy of energetic reserves and oxygen. Moreover, and more important, it reduces to a minimum the production of carbon dioxide, which is a major factor when the animal is placed in a confined space. Probably, the conditions are comparable for a hibernating homeotherm. To investigate this we shall examine first what happens when a homeotherm is placed in an artificial state of hibernation.

5. Modifications of Thermogenesis by Chlorpromazine (Largactil)

Chlorpromazine, or Largactil, or chloro 3 (dimethylamino-3′ propyl)—10 phenothiazine, has the formula:

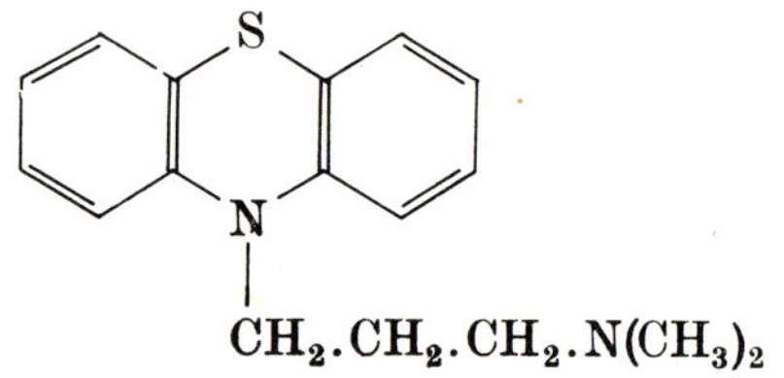

It possesses multiple physiological actions: hypo-thermizing, hypnotic, inhibitor of conditional reflexes, anti-emetic, anti-diuretic, and also potentializing many analgesic and anaesthetic drugs. It has been studied by Courvoisier, Decourt, Laborit, and Huguenard, and used for many medical purposes, ranging from artificial hibernation in surgery to tranquilizing in phsychiatry. The following account will deal chiefly with two of its effects: firstly its ability to depress and regularize thermogenesis, and secondly its potentializing action on anaesthetics.

Figure 16.7 shows, at I, the effects produced on the thermogram of a one day old mouse by a dose of 0.8 mg of chlorhydrate of chlorpromazine per gram (administered orally), at 37.1°. For 9 hr the thermogenesis was perfectly regularized on a low level (AB_1); then, after the 9th hour, small undulations B_1 B_2 B_3 B_4 appeared. During the 12th hour, larger undulations $C_1\, C_2\, C_3\, C_4$ . . . . intervened, together with a rise in the average flow, and the behaviour returned to normal. This curve is thus a recording of the "thermogenetic awak-

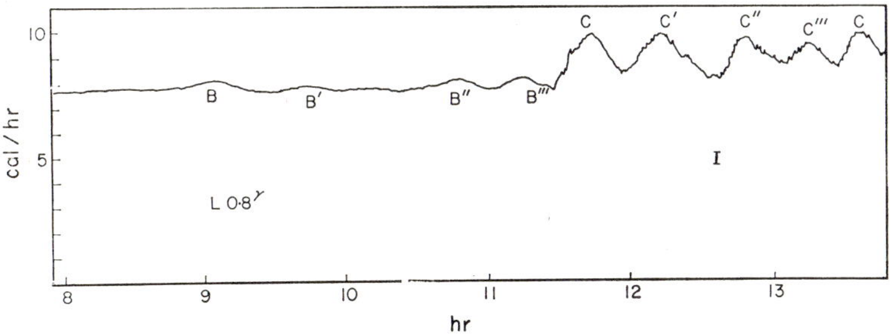

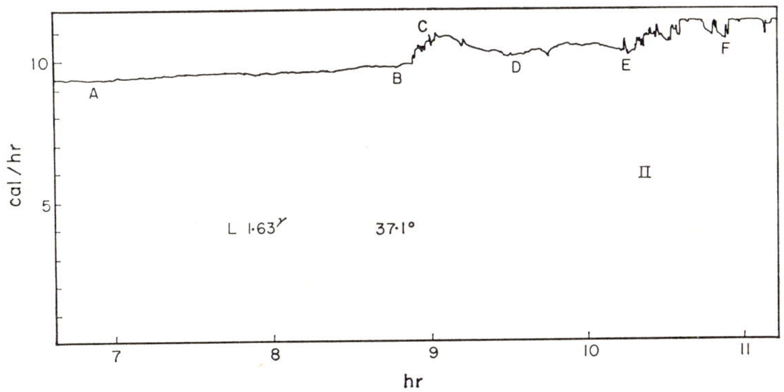

FIG. 16.7. Effect of Largactil (chlorpromazine), given orally, on the thermogenesis of young mice (24—48 hr) at 37.1°C. I, Dose of 0.8 microgram per gram: $B_1B_2B_3$ B_4: beginning of the thermogenetic awakening, after 9 hours of complete uniformity. II, Dose of 1.63 microgram per gram. *BCD:* thermogenetic awakening.

ening" of the animal, not to be confused with the anaesthetic awakening of the previous cases, hypothermy being a different phenomenon from anaesthesy.

Curve II, Fig. 16.7, taken at the same temperature, shows the effect of a stronger dose of chlorpromazine: 1.63 mg per gram, on an older animal (48 hr). After a period of perfect regularization *AB*, lasting for

9 hr, the thermogenetic awakening occurred at *BCD*, with a shorter intermediate period (*BE*) than in the previous case. The normal thermogenesis *EF* was reached after 11 hr and the awakening phase extended over one hour and a half instead of 3 hr.

Figure 16.8 shows the potentializing action of chlorpromazine on the anaesthesy by Nembutal at 25°. In both cases *AB* is the normal thermogram. For the curve I, at *B*, the sub-

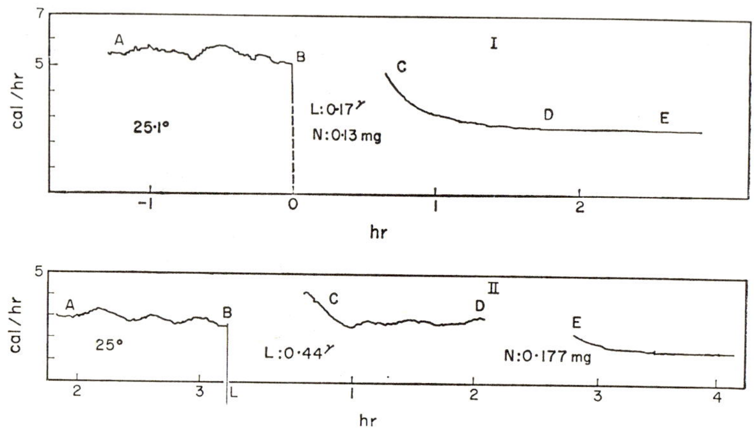

FIG. 16.8. Synergetic effect of Largactil and Nembutal on the thermogenesis of young mice (12—24 hr) at 25°C. *A B* normal thermogram. I, Doses of 0.17 microgram of Largactil and 0.13 milligram of Nembutal per gram, given simultaneously. *C D E* rapid depression and complete regularization of the heat flow. II, *C D:* 0.44 microgram of Largactil given first: no effect; *E:* two hours later 0.1777 mg of Nembutal is also given: depression and regularization.

ject was removed from the calorimeter and given orally a dose of 0.17 microgram of Largactil per gram, together with 0.13 milligram of Nembutal per gram. These doses are too small to produce any noticeable effect if administered separately. But, jointly, they produced a rapid depression of thermogenesis *(CDE)*, with a perfect uniformization of the flux. This state of low and regular thermogenesis persisted for more than 20 hr without interruption; its level *DE* is about 45% of the normal one *AB*.

In curve II the two drugs were given successively: first a dose of 0.44 microgram of Largactil per gram, then, 2 hr later, a dose of 0.177 milligram of Nembutal per gram. It is obvious that the first administration alone exerted no effect: the resulting thermogram *CD* is exactly the same as the normal one *AB*. But the addition of the second drug produced at *EF* a depression and uniformization of the flux, as previously observed. If the drugs were administered in the reverse order (Nembutal first) the same result was recorded, i. e. no effect for the first dose, and a strong depression and regularization for the second. This phenomenon is called the synergetic effect or potentialization.

6. Hibernating Mammals

The recording microcalorimeters provide a valuable means of studying the phenomenon of hibernation in warm-blooded animals. The 18 mm cells of the early models were too small for these experiments but, with the construction of cells of 35 mm diameter, we have been able to carry out hibernation studies. Up to now our subjects for this purpose have been bats *(Myotis subulatus)*. When they are hibernating, i.e. asleep and entirely motionless and insensible, being kept at low temperature (10°C), their thermogenesis is low and uniform, comparable to that displayed by cold blooded animals at the maintenance level. When warmed to 20° or more, they awake progressively: their thermogenesis rises slowly, first with some faint undulations such as shown in Fig. 16.7. After some hours the animal regains a higher level of thermogenesis with fluctuations; the internal body temperature rises: the animal starts to move and to react to external excitations, finally regaining its normal level of activity. This is a case of thermogenetic awakening, similar to the one we have described above after an artificial hibernation due to chlorpromazine. Such recordings afford new means of studying the phenomenon of hibernation. It seems probable that an unknown hormone is released which influences the mechanism of thermoregulation in a manner similar to that by chlorpromazine, but it remains to isolate this agent and to find out how it is produced and released. Why some mammals, when winter approaches, react in this way, whilst others, on the contrary, increase their thermogenesis and metabolism, is very much an open question and one of the most puzzling in the entire field of comparative physiology.

7. Influence of Curarizing Drugs on Thermogenesis

On vertebrates the curarizing drugs possess a strong depressive action on thermogenesis. Figure 16.9 shows the influence of an injection of 0.8 microgram of tubo-curarin

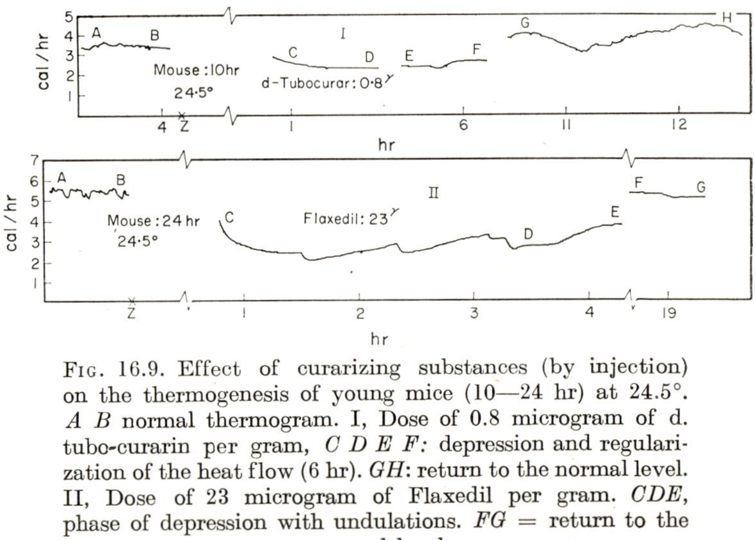

Fig. 16.9. Effect of curarizing substances (by injection) on the thermogenesis of young mice (10—24 hr) at 24.5°. *A B* normal thermogram. I, Dose of 0.8 microgram of d. tubo-curarin per gram, *C D E F:* depression and regularization of the heat flow (6 hr). *GH*: return to the normal level. II, Dose of 23 microgram of Flaxedil per gram. *CDE*, phase of depression with undulations. *FG* = return to the normal level.

(natural curare) per gram on a ten hours old mouse. Thermogram *CD* shows a neat depression below the normal level *AB*, with a regularization of the thermic flow lasting five hours. During the 6th hour the thermogenesis rises again, with reappearance of some undulations (*EF*). After 11 hr the normal level is reached again and even surpassed (*GH*), with undulations larger than usual.

We have compared this thermogenetic action of natural curare with that of artificial drugs possessing similar physiological effects. One of these is triido-ethylate of gallamin (Flaxedil or 3697 RP), viz:

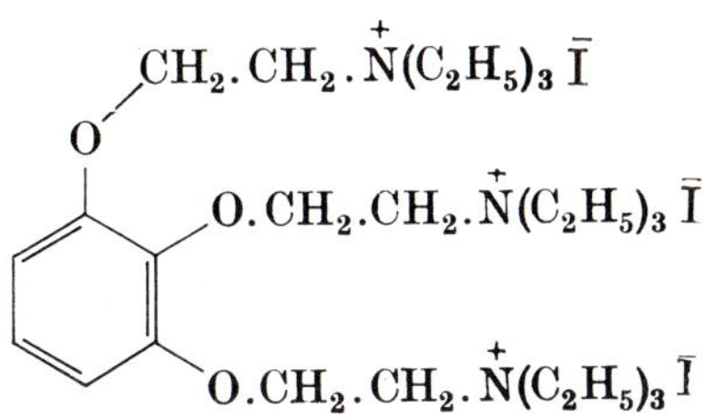

Curve II, Fig. 16.9, shows the results of injection of 23 micrograms of this substance per gram on a one day old mouse. The physiological effect is a curarization, i.e. entire muscular palsy without loss of sensitivity. From the thermogenetic point of view it may be noticed that the depression *CD*, as compared to the normal level *AB*, is greater than in the preceding case, but that the thermic flow is not here regularized: marked undulations are apparent at *CD*. 19 hr later, the normal level is reached at *FG*, but the usual small undulations of *AB* have not yet reappeared.

Curve I of Fig. 16.10 shows the effect of the same substance on an adult mouse, with a smaller dose (19 micrograms per gram). Here the regularization is not reached: small undulations persist at *CD*, with marked paroxysms *E* after 12 hr. But the depression remains considerable by comparison with the normal level *AB*.

Another interesting curarizing drug is the suxamethonium bromide (Brevidil),

$$\begin{array}{l} CH_2.COO.CH_2.CH_2.\overset{+}{N}(CH_3)_3\overline{Br} \\ | \\ CH_2.COO.CH_2.CH_2.\overset{+}{N}(CH_3)_3\overline{Br} \end{array}$$

Its action is strong but less lasting than that of Flaxedil. Curve II, Fig. 16.10 shows the result of an injection of 17 micrograms of Brevidil per gram on a mouse 6 hr old. Compared with the normal level *AB*, the depression of thermogenesis is strong (A_1 B_1) but short, lasting only three hours. Then, the thermic flow rises again (B_1 C), becoming, at the 5th hour, even higher than normal (at C). After some hours, however, the usual level is regained.

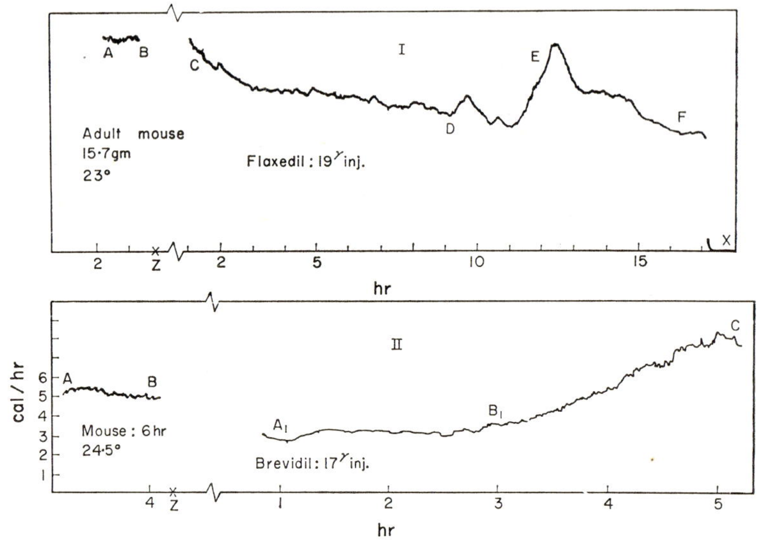

FIG. 16.10. Effect of curarizing substances (by injection) on the thermogenesis of mice. AB = normal thermogram. I, Dose of 19 microgram of Flaxedil per gram on an adult male, at 23°, CD = depression, E = temporary paroxysm. II, Dose of 17 microgram of Brevidil per gram, on a newborn (6 hr old at the beginning of the experiment), at 24.5°. A_1B_1 = temporary depression of the heat flow. B_1C = secondary rise, above the normal level AB.

Those examples are sufficient to show that the recording microcalorimeters are perfect tools for comparing the efficiency of curarizing drugs and for studying their mode of action.

8. Influence of Spasmolytic Derivatives of Phenothiazine on Thermogenesis

The hibernating effects of chlorpromazin have been described above (section 6). Other substances of the same chemical family, (i.e. derivatives of phenothiazine), possess various other physiological actions, namely spasmolytic, anti-emetic, anti-histaminic, neurolytic etc. They are becoming increasingly used in medicine for the treatment of sea sickness and allergies. Here we shall examine them only from the point of view of their thermogenetic action.

The formula of prochlorperazin (Stemetil, or 6140 RP) is:

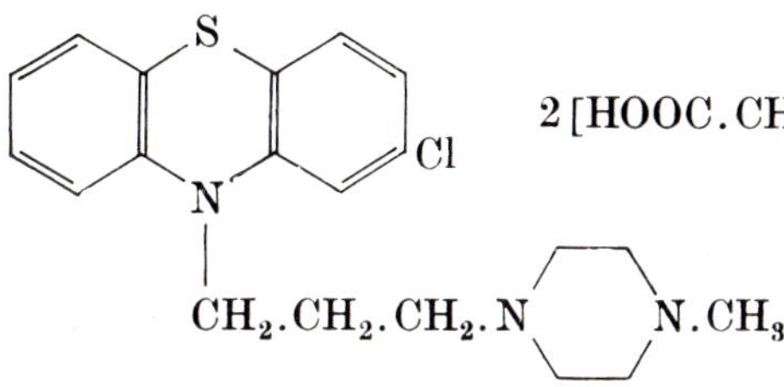

$2[HOOC.CH{=}CH.COOH]$

Courvoisier, Ducrot, Fournel and Julou (1957a) have shown that its anti-emetic action is four times higher than that of chlorpromazin, but that its hypothermizing influence is smaller, as also is its potientializing action toward anaesthetics. Figure 16.11 shows at I the influence of a dose of 40 micrograms of Stemetil per gram administered by mouth to an adult mouse, at 22°. The hypothermizing action is here obvious, with a depression (*CDE*) of about 50% of the thermogenesis (*AB* being the normal level), and an almost perfect regularization of the flux lasting more than 15 hr (Prat, 1959). It is interesting to notice that the toxicity of this compound is very low, the ratio between the lowest active dose and the first toxic dose being 1 : 3000. This is in marked contrast to tubo-curarin, for which the margin is very narrow.

Curve II, Fig. 16.11 shows the action of another phenothiazinic compound: aminopromazine (Lispamol or 3828 RP); here the dose is 50 micrograms per gram and the age of the subject, 6 hr. The result is only a depression of the thermic flux (*CD*) without regularization. This drug is known as a spasmolytic and an anti-histaminic. Its thermogenetic activity is obviously smaller.

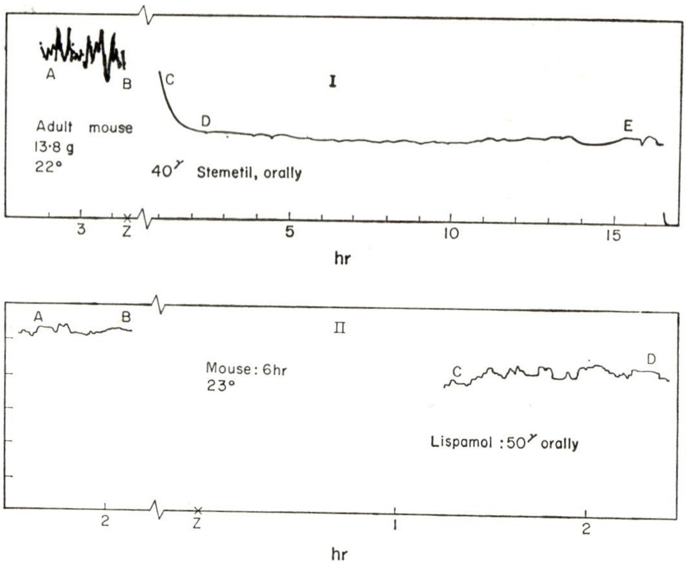

FIG. 16.11. Effect of phenothiazine compounds (given orally) on the thermogenesis of mice, at 22—23°. *AB* normal thermogram. I, Dose of 40 microgram of Stemetil per gram on an adult male; *CDE* = depression and regularization of the thermic flow. II, Dose of 50 microgram of Lispamol per gram of a new-born (6 hr); *CD* = depression without regularization.

A recently prepared compound is noteworthy because of its very high activity; this is levomepromazine (Nosinan of 7044 RP), viz.:

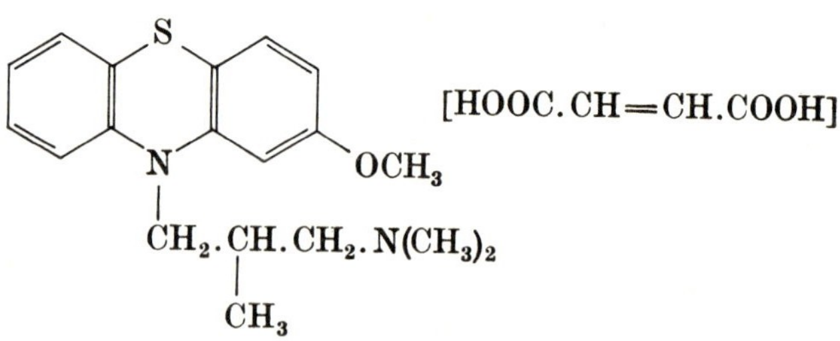

Courvoisier, Ducrot, Fournel and Julou (1957b) have demonstrated the neuroleptic, spasmolytic and anti-histaminic properties of this drug, as well as its potentializing effects on narcosis. Sigwald, Henne, Bouttier, Raymondeaud and Quetin (1956) have described its applications in neurology and psychiatry.

Curves I and II, Fig. 16.12 show the thermogenetic action. In curve I the dose was 0.9 microgram of levomepromazine per gram, given

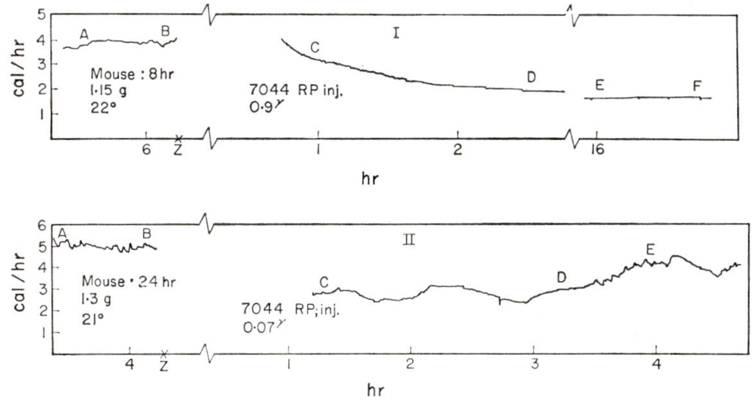

FIG. 16.12. Effect of Levomepromazine (Nozinan or 7044 RP), by injection, on the thermogenesis of young mice (8 hr—24 hr), at 21—22°. *AB* = normal thermogram. I, Dose of 0.9 microgram per gram. *CDEF* = depression with regularization, persisting more than 16 hr. II, Dose of 0.07 per gram; *CD* = depression with undulations; *E* = return towards normal.

to a mouse 8 hr old. Starting from the normal level *AB*, the depression *CD* of the thermic flux is obvious, as well as a regularization lasting more than 16 hr (*EF*).

In curve II, Fig. 16.12 the dose was even smaller: 0.07 microgram per gram, i.e. a concentration of only 7×10^{-8} in the body of the animal. Nevertheless, there was an obvious reaction, the thermic flux dropping (*CD*) to less than half the normal level (*AB*). On this occasion, however, no regularization occurred, large undulations persisted (*CD*), and the action was shorter: after 3 hr the flux rose again (*DE*) and reached the normal level after 4 hr.

By comparing Figs 16.12, 16.9 and 16.7, it is clear that levomepromazine is even more active on thermogenesis than tubo-curarin and Largactil.

CHAPTER 17

COMPARISON OF THERMOGENETIC SOURCES

We have presented a variety of thermograms from plants, micro-organisms and animals, and drawn attention to the specific characteristics and differences in shape and form of the curves. For the general purpose of comparison, we shall consider the *average* heat output given by a particular category of living being: in this way, it is possible to present the comparison in the form of tables. The table in Fig. 17.1 compares the *total* heat output from each individual category, that in Fig. 17.2 relates to the heat output *per gram weight* of each individual category.

The scale of heat output in Fig. 17.1 ranges from 10^{-9} cal/hr, by units of ten, to 10^6 cal/hr. At the bottom of the scale, liberating 10^{-9} cal/hr, we find the *coli bacillus*, during the process of division. The average heat output of a human cell places it two stages higher in the scale at 10^{-7} cal/hr. The output from a grain of wheat in germination, and for a drosophila in flight, is of the order 0.03 cal/hr. The thermogenesis from a grasshopper amounts to several cal/hr, from a young mouse several tens of cal/hr, and from an adult mouse several hundreds of cal/hr. Finally, to complete the table, we have included the dog (10^5 cal/hr), and man, whose output varies from 10^5 cal/hr whilst at rest, to 5×10^5 cal/hr during periods of intense physical effort, lasting for a few seconds (Winslow and Herrington, 1949).

The order in Fig. 17.1 is completely altered when we classify categories of living being in terms of heat output *per gram weight.* As shown in Fig. 17.2, it is now the bacterial cell which heads the table, liberating 400 cal/hr/g, followed by the drosophila at from 10–40 cal/hr/g, man at 1–4 cal/hr/g, and the wheat grain at from 0.1–0.4 cal/hr/g.

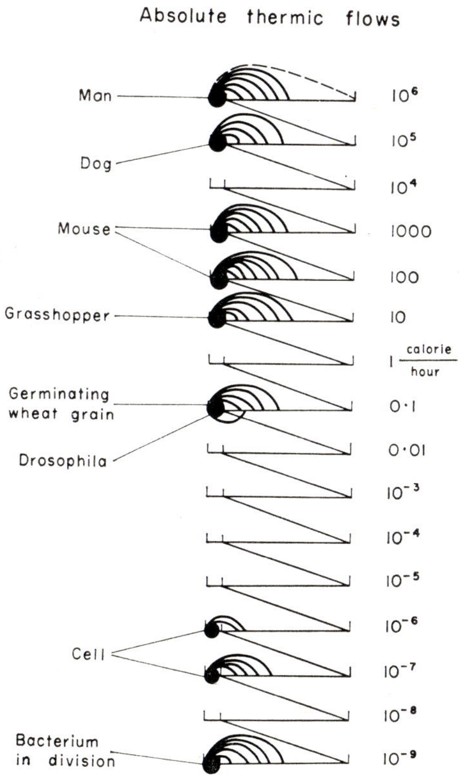

FIG. 17.1. Scale showing total heat outputs (Calvet and Prat, 1956).

On the same scale, the heat output from the sun is well below that of any living being: astrophysicists have estimated that the sun radiates *ca.* 4×10^{29} cal/hr, which, in relation to its huge mass of 2×10^{33} g, represents only a very modest output of 2×10^{-4} cal/hr/g – a million times less than the corresponding output of a bacterium (Calvet and Prat, 1956).

This fact seems, at first sight, almost unbelievable: but, of course, one must bear in mind that the enormous size of the sun gives it a very small surface area relative to mass. Thus

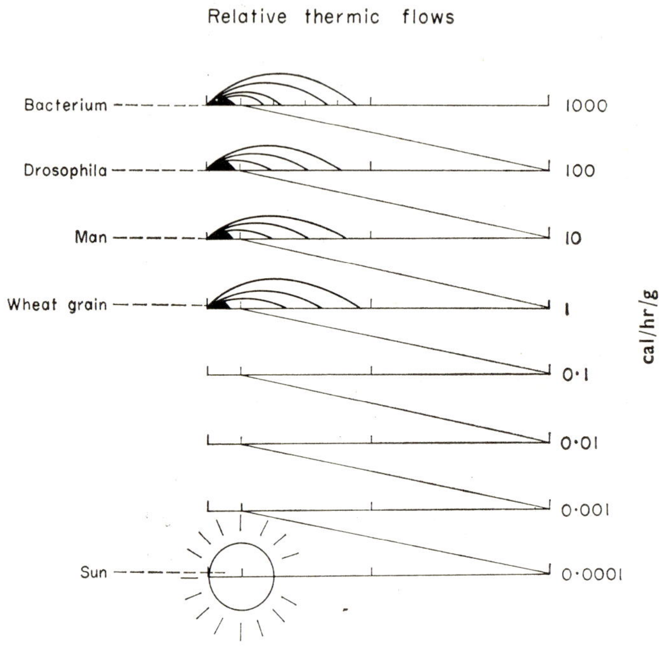

FIG. 17.2. Scale of heat outputs per gram of material (Calvet and Prat, 1956).

although the intensity of radiation per unit surface area of the sun is enormous, the heat produced per gram of solar material inside the star is very small. Nevertheless, the table given in Fig. 17.2 presents a striking feature of living matter, especially of micro-organisms, and offers scope for deeper speculation.

Conclusion

The advent of the continuous recording microcalorimeter has opened up extensive and varied possibilities for thermal studies in the physical, chemical, and biological sciences. The microcalorimeter, reproducible in performance and sensitive in response, is capable of detecting the smallest changes occurring during a chemical reaction, during the germination of a seed, in the growth of a bacterial culture during the metamorphosis of an animal, etc.

Microcalorimetric studies can be planned in a systematic manner: one can, for example, study a single material or process under variable conditions, altering in turn temperature, humidity, atmospheric composition, or solution composition; or one can investigate a variety of materials or processes under identical conditions. The analysis of the collected data from such systematic studies provides insight into the mechanism of reactions, the comparative physiology of animals, plants, and micro-organisms, and informs on the effects of external agents on living matter.

We have shown how the microcalorimeter can be used to measure heats of solution, of mixing, of dilution, of gelatinization, and for studies on alumina gels, nitrocelluloses, and cements. In agronomy, microcalorimetric studies can be used in testing seeds, and in studying the effects of herbicides, fertilizers, and growth-hormones on plant cultivation. In microbiology, microcalorimeters will become increasingly useful for comparing bacterial strains, and for testing antibiotics; in animal physiology, microcalorimetry has its place in the study of pharmacodynamics, in toxicology, and in analysing the process of aging.

The potential applications of the microcalorimeter in experimental science are almost numberless, and we have only been able to give a brief account of this new instrument and its applications in this short book. For those readers wishing to delve deeper into the subject, we have included an extensive bibliography, and might also refer them to the more detailed treatise on microcalorimetry (Calvet and Prat, *Microcalorimétrie; applications physicochimiques et biologiques,* Masson, 1956), recently published.

BIBLIOGRAPHY OF PART 3

1911 A. V. Hill; A new type of differential microcalorimeter for the estimation of heat produced in physiological, bacteriological or fermentation action. *J. Physiology,* **43,** 316.

1927–8 W. K. Slater; The aerobic and anaerobic metabolism of the common cockroach. *Biochem. J.*, **21,** 198; **22,** 331.

1929 S. Bayne-Jones and H. S. Rhees; Bacterial calorimetry, *J. Bacteriology,* **17,** 123.

1933 C. F. Schmidt and S. Bayne-Jones; Bacterial calorimetry. *J. Cellular Comp. Physiol.,* **3,** 405.

1937 I. R. Taylor and F. Crescitelli; Measurement of heat production. *J. Cellular Comp. Physiol.*, **10,** 93.

1938 R. Pratt; The influence of indole-3-acetic acid on respiration and growth of wheat seedlings. *Amer. J. Botany,* **25,** 389.

1939 C. E. A. Winslow and H. H. Walker; The earlier phases of the bacterial culture cycle. *Bacteriol. Review,* **3,** 147.

V. B. Wrigglesworth; *The principles of insect physiology.*

1941 A. Krogh and E. Zeuthen; The mechanism of flight preparation in some insects. *J. Exper. Biol.*, **18,** 1.

1944 (a) H. Prat and E. Calvet; The application of microcalorimetry to analysis of the initial stages of germination. *C. R. Soc. Biol.*, **138,** 662.

(b) H. Prat; Recognition of a "hydro-state" and the influence of external factors during germination. *C. R. Soc. Biol.*, **138,** 664.

(c) H. Prat; The influence of indol-acetic acid and of alcohol on germination. *C. R. Soc. Biol.*, **138,** 675.

(d) H. Prat and E. Calvet; The evolution of heat during the hydration of seeds. *C. R. Acad. Sci., Paris*, **218, 944.**

1945 (a) H. Prat and E. Calvet; The effect of dehydration of seeds on the thermogenesis of germination. *C. R. Acad. Sci., Paris,* **220,** 226

(b) H. Prat and E. Calvet; The use of thermograms to characterize specific properties of seeds. *C. R. Acad. Sci., Paris,* **220,** 470.

(c) H. PRAT and E. CALVET; Modifications in the thermogenesis of germination resulting from drying of seeds. *C. R. Soc. Biol.*, **139**, 222.

(d) H. PRAT; Histo-physiological gradients in relation to vegetable growth. *Rev. Can. Biol.*, **4**, 543.

E. CALVET, J. FRICKER and H. PRAT; Measurements of the thermogenesis of bacteria. *C. R. Acad. Sci., Paris*, **220**, 7, 97 and 798.

1946 L. CHAMBOST and E. CALVET; The thermogenesis of poïkilotherms. *C. R. Acad. Sci., Paris*, **223**, 51.

H. PRAT, E. CALVET and J. FRICKER; Microcalorimetric studies on the growth of bacteria and the germination of seeds. *Rev. Can. Biol.*, **5**, 247.

J. SIGWALD, D. BOVET and G. DUMONT; The treatment of Parkinson's disease by the hydrochloride of diethylamino-N-ethylthio diphenylamine. *Rev. Neurologique*, **78**, 581.

1947 H. PRAT; Variations of thermogenesis in the early stages of germination. *Trans. Roy. Soc. Can.*, **V 41 (3)**, 95.

1948 B. N. HALPERN, G. PERRIN and P. DEWS; The strength of some synthetic antihistamines as local anaesthetics. *J. Physiologie*, **40**, 210.

H. PRAT; Histo-physiological gradients and plant organogenesis. *Bot. Review*, **14**, 603; **17**, 693.

C. WINTER; The potentializing effect of antihistamine drugs on the sedative action of barbiturates. *J. Pharm.* and *Expt. Therapeutics*, **94**, 7.

1949 P. CHAUVIN; The physiology of the insect. *Inst. Nat. Rech. Agron., Paris.*

C. E. A. WINSLOW and L. P. HERRINGTON; *Temperature and Human Life.*

1950 H. LABORIT; Potentialization of general anaesthetics: (and further notes): *Presse medicale*, **58**, 416 and 851; **59**, 1329; **60**, 206 and 347.

1951 D. BOVET, E. BOVET-NITTI, S. GUARINO, V. B. LONGO and R. FUSCO; Studies on synthetic curarizing drugs. *Arch. Int. Pharmacodyn.*, **88**, 1.

P. CHOUARD; Dormancy and inhibitions of seeds. *Centre Doc. Univ.*

(a) S. and A. GELINEO; Oxidation in rats in first days after birth. *Bull. Acad. Serbe des Sc.*, III. *Sc. Medic.*, 119.

(b) S. and A. GELINEO; Environmental temperature of the nest, and the appearance of chemical thermoregulation in rats at 21°C. *Bull. Acad. Serbe des Sc.*, III. *Sc. Medic.*, 149.

(c) S. and A. GELINEO; Thermoregulation in a new born rat at the nest temperature. *C. R. Acad. Sci., Paris* **232**, 1031.

H. PRAT; Influence of surrounding temperature on the thermogenesis of germination. *Rev. Can. Biol.*, **10**, 443.

L. M. ROTH and E. H. WILLIS; Hygroreceptors in Coleoptera, *J. Expt. Zool.*, **117**, 451.

1952 L. V. HEILBRUNN; *An outline of general physiology.*

M. P. HUGUENARD; Artificial hibernation — new and useful data. *Semaine des Hôpitaux*, **33**, 1407.

H. LABORIT; From neuroplegia to artificial hibernation. *Gazette Medic.*, **59**, 1423.

H. PRAT; Microcalorimetric studies on the germination of cereals. *Can. J. Bot.*, **30**, 379.

1953 S. COURVOISIER, J. FOURNEL, R. DUCROT, M. KOLSKY and P. KOETSCHET; Pharmacodynamic properties of the hydrochloride of chloro-3(dimethylamino-3′ propyl)-10 phenothiazine. *Arch. Int. Pharmacodyn. XCII*, 305 (3–4).

P. DECOURT; The therapeutic action of chlorpromazine. *Therapie*, VII, 846.

L. PERUZZO and R. B. FORNI; Action of chlorpromazine on the respiration of tissues. *Presse medicale*, **61**, 1463.

H. PRAT; Observations on bacterial thermogenesis. *Rev. Can. Biol.*, **12**, 19.

K. D. ROEDER; *Insect Physiology*, Wiley, New York.

1954 R. G. BUSNEL and G. OBOLENSKY; Effect of ultra-sonic vibrations on the speed of germination and the growth of barley. *C. R. Acad. Sci., Paris.*

P. DECOURT; Narcobiotic therapeutics and artifial hibernation. *Bull. med.*, **62**, 855.

J. GIAJA and L. MARKOVIC-GIAJA; Chlorpromazine and thermoregulation. *C. R. Soc. Biol.*, **148**, 842.

H. LABORIT and M. P. HUGUENARD; *Pratique de l'hibernotherapie*, Masson, Paris.

M. LAMARCHE and P. ARNOULD; Action of chlorpromazine on the intestine (isolated) of the rabbit and the rat. *C. R. Soc. Biol.*, **148**, 565.

V. POPOVIC; Chlorpromazine and induced poïkilothermy. *C. R. Soc. Biol.*, **148**, 845.

H. PRAT; Microcalorimetric analysis of the variations in thermogenesis in different insects. *Can. J. Zool.*, **32**, 172.

H. PRAT; Thermogenesis of the new-born mouse. *Rev. Can. Biol.*, **13**, 18.

P. VIAUD; Amines derived from phenothiazine. *J. Pharmacy and Pharmacology*, **6**, 361.

1955 R. G. GRENELL, J. MENDELSON and W. D. MCELROY; Effects of chlorpromazine on the metabolism in central nervous systems. *Arch. Neurology and Psychiatry*, **73**, 347.

H. PRAT; The influence of anaesthetics on the thermogenesis of a young mouse. *Rev. Can. Biol.*, **14**, 1.

1956 (a) H. PRAT; Thermogenesis of *Periplaneta americana:* effects of olfactory excitations and decapitation. *Rev. Can. Biol.*, **15**.

(b) H. PRAT; Thermogenesis of *Periplaneta americana. 10th Congress. Int. Entomol: Proceedings* **2**, 279.

J. SIGWALD, M. HENNE, D. BOUTTIER, C. RAYMONDEAUD and A. QUETIN; The activity of a new phenothiazine derivative in psychiatry and neurology. *Presse Medicale,* **87**, 2011.

1957 J. CHEYMOL; Qualitative and quantitative assay of curarizing substances. *Therapie,* **12**, 321.

S. COURVOISIER, R. DUCROT, J. FOURNEL AND L. JULOU; Pharmacodynamic properties of prochlorperazine, and levomepromazine. *C. R. Soc. Biol.*, 1144 and 1378.

G. OBOLENSKY; The stimulation of plants by ultra-sonic vibrations. Thèse Paris.

H. PRAT; Microcalorimetric analysis of the effects of Largactil, Phenergan and Nembutal on the thermogenesis of young mice. *Rev. Can. Biol.*, **15**, 336.

A. PUNT, W. J. PARSER and J. KUCHLEIN; Oxygen uptake in insects with cyclic CO_2 release. *Biol, Bull.*, **112**, 108.

1958 H. PRAT; Action of curarizing drugs on the thermogenesis of animals. *Therapie,* **13**, 833.

1959 H. PRAT; Microcalorimetric analysis of the effect of phenothiazine derivatives on the thermogenesis of mice. *Therapie,* **14**, 594.

1960 H. PRAT and L. ROBERGE; Changes in thermogenesis with aging of mice. *Rev. Can. Biol.*, **19**, 80.

H. PRAT; A natural classification of graminaceae. *Bull. Soc. Bot. Fr.*, **107**, 32.

1962 H. PRAT; Biological and Zoological Thermogenesis, Chap. 18, *Experimental Thermochemistry,* Vol. 2, Interscience Wiley, New York.

NAME INDEX

SUBJECT INDEX